Kakuro Puzzle Books
Cross Products
Easy

200 Mind Teasers Puzzle
Large Print
Book 10

This Book Belongs To:

_____

Copyright and other intellectual property laws protect these materials. Reproduction or retransmission of the materials in whole or part thereof in any manner, without the prior written consent of the copyright holder, is a violation of the copyright law and will be enforced to the fullest extent of the applicable law.

Puzzles are created using Crossword Express.

Copyright © 2019 Panda Puzzle Book
All Rights Reserved

What is Kakuro Cross Products?

A KAKURO puzzle is constructed on a crossword grid just like a standard crossword, but the digits 1 to 9 are used instead of the letters of the alphabet. In a standard Kakuro puzzle, the Across and Down clues are simply the sums of the digits in the across and down words.

A variant of Kakuro is *Cross Products* (or *Cross Multiplication*), where the clues are the product of the digits in the entries rather than the sum.

The rules of Kakuro are simple:

1. Each cell can contain numbers from 1 through 9

2. The clues in the grey cells tells the multiplication of the numbers next to that clue. (on the right or down)

3. The numbers in consecutive white cells must be unique. No number may be used in the same block more than once.

Example

Puzzle   Solution

Puzzle 1

	63	7
63 / 15		
21		
5		■

Puzzle 2

	12	32
48 / 42		
48		
7		■

Puzzle 3

	12	72
54 / 4		
32		
2		■

Puzzle 4

	168	8
12 / 7		
112		
7		■

Puzzle 5

20
18
36
36
18
20

Puzzle 6

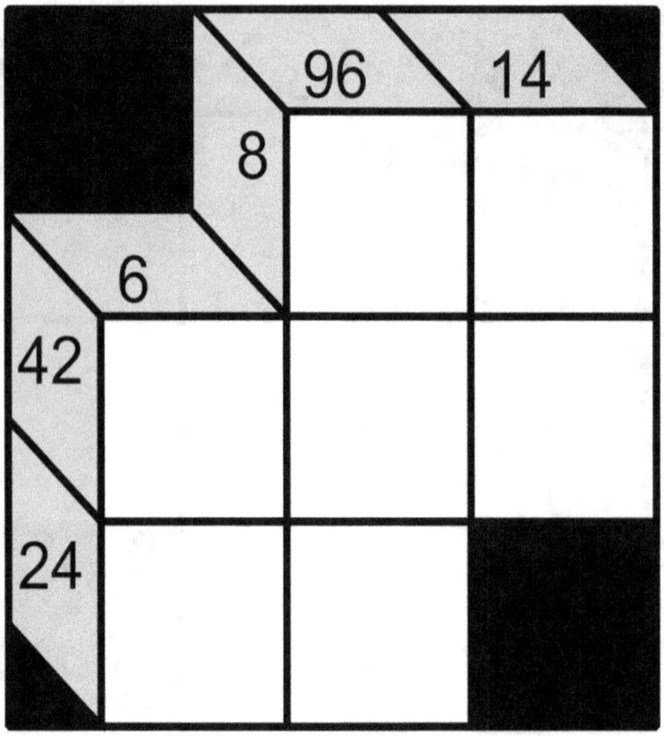
96
14
8
6
42
24

Puzzle 7

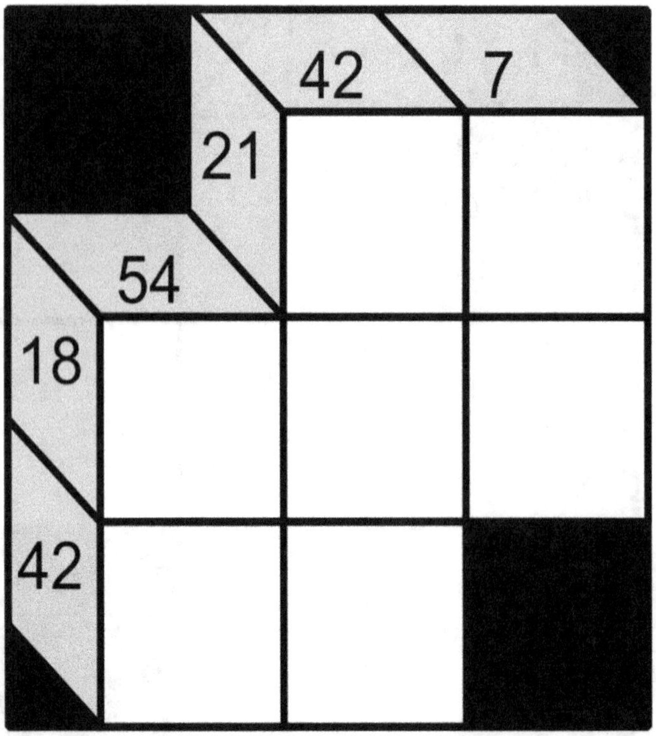
42
7
21
54
18
42

Puzzle 8

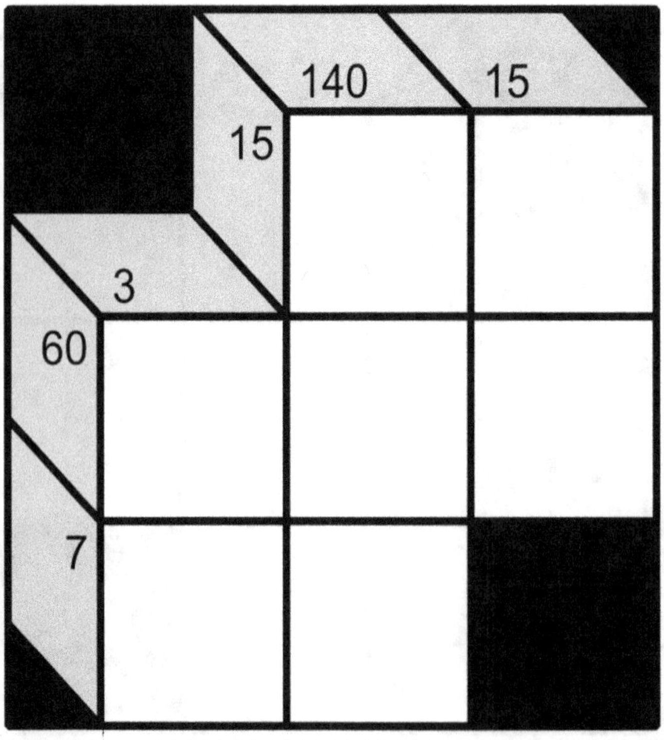
140
15
15
3
60
7

Puzzle 9

	360	18
54		
7		
168		
5		

Puzzle 10

	378	8
18		
36		
216		
28		

Puzzle 11

	270	36
45		
24		
288		
18		

Puzzle 12

	96	40
32		
28		
60		
56		

Puzzle 13

	48	21
56		
20		
90		
4		

(clues: 56, 20, 90, 4 on left; 48, 21 on top)

Puzzle 14

	42	27
21		
8		
108		
4		

Puzzle 15

	15	16
2		
56		
168		
40		

Puzzle 16

	20	7
7		
54		
45		
24		

Puzzle 17

70
14
14
8
70
8

Puzzle 18

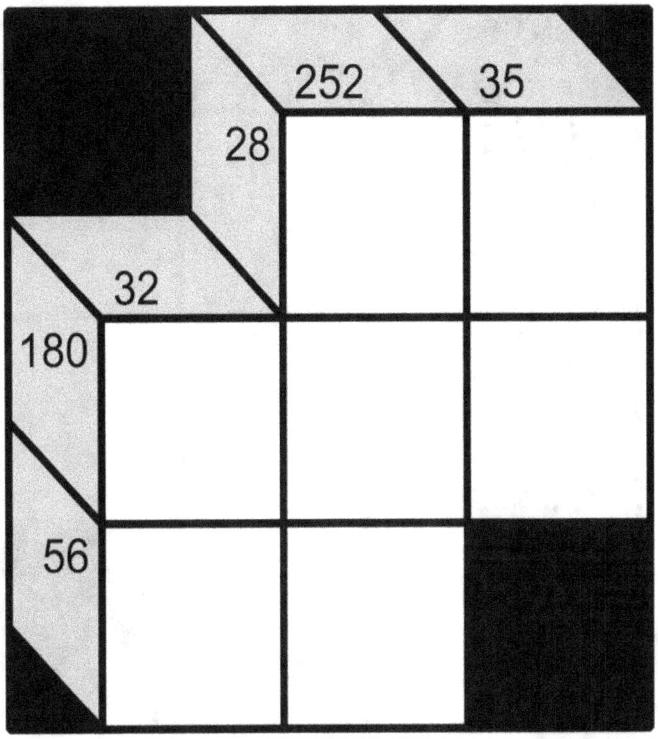
252
35
28
32
180
56

Puzzle 19

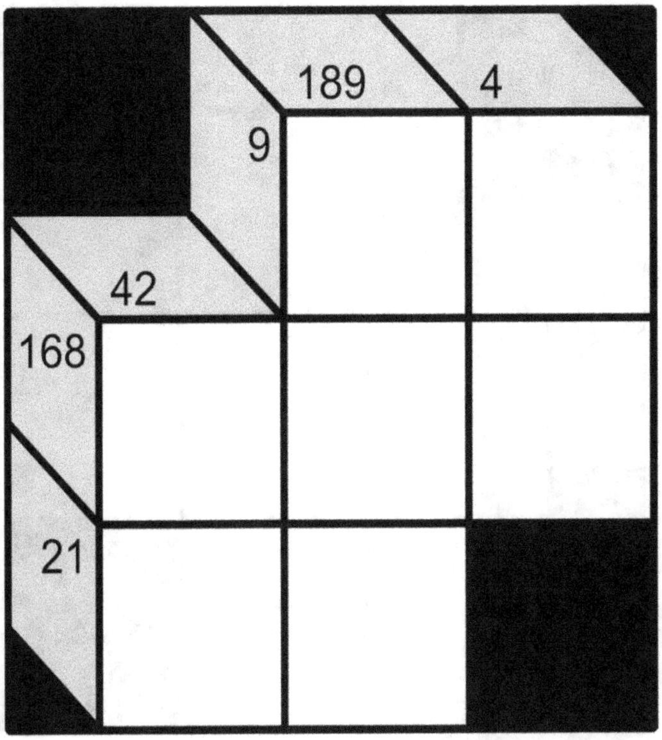
189
4
9
42
168
21

Puzzle 20

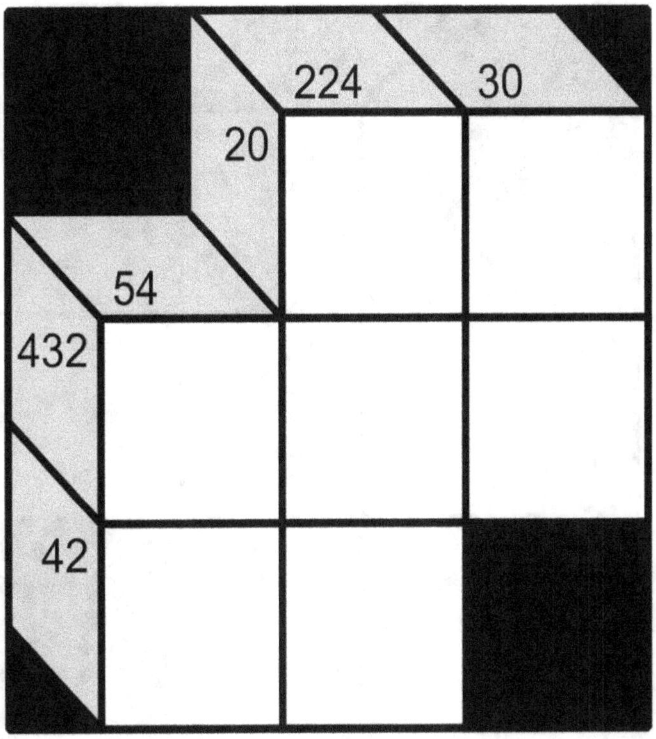
224
30
20
54
432
42

Puzzle 21

Puzzle 22

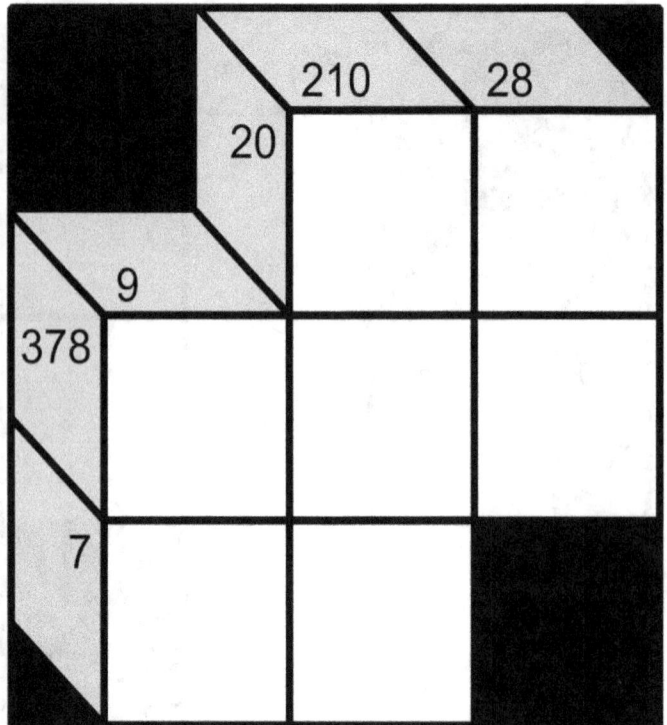

Puzzle 23

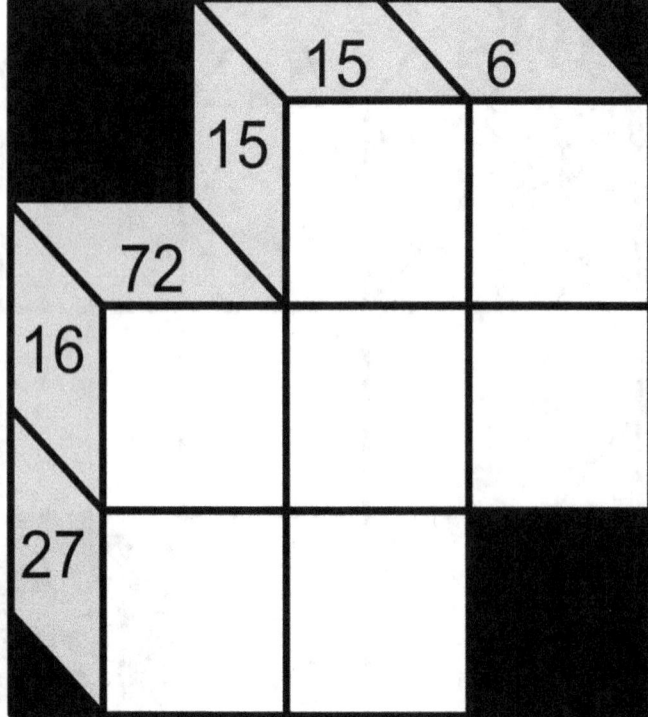

Puzzle 24

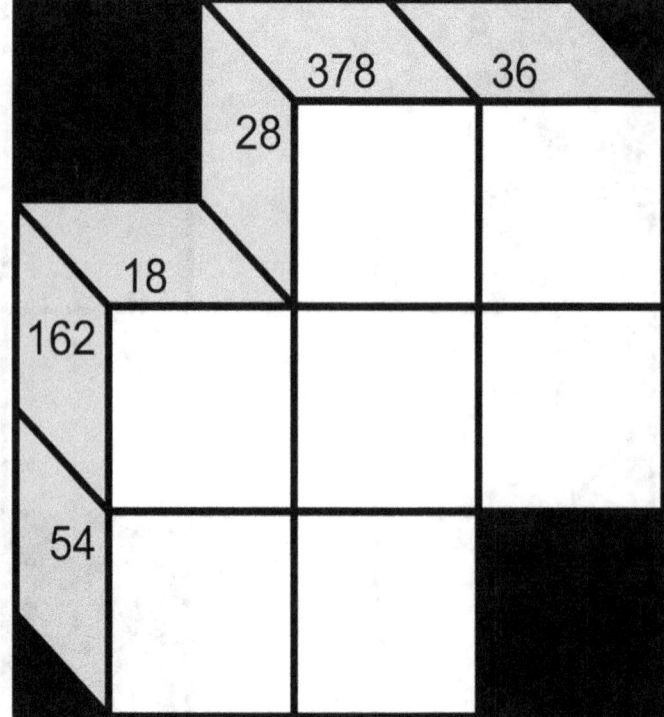

Puzzle 25

	42	35
21 / 36		
40		
63		■

Puzzle 26

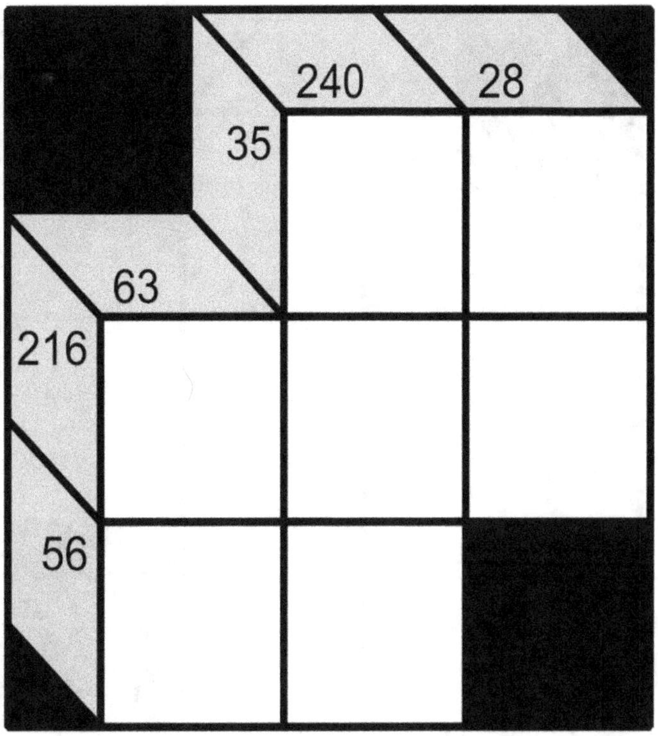

	240	28
35 / 63		
216		
56		■

Puzzle 27

	84	14
28 / 24		
84		
12		■

Puzzle 28

	432	6
48 / 20		
36		
30		■

Puzzle 29

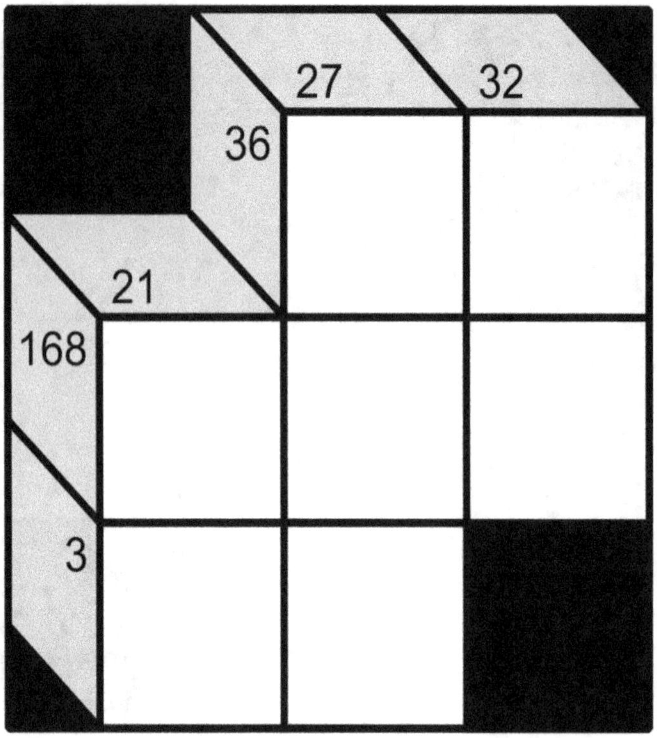

Puzzle 30

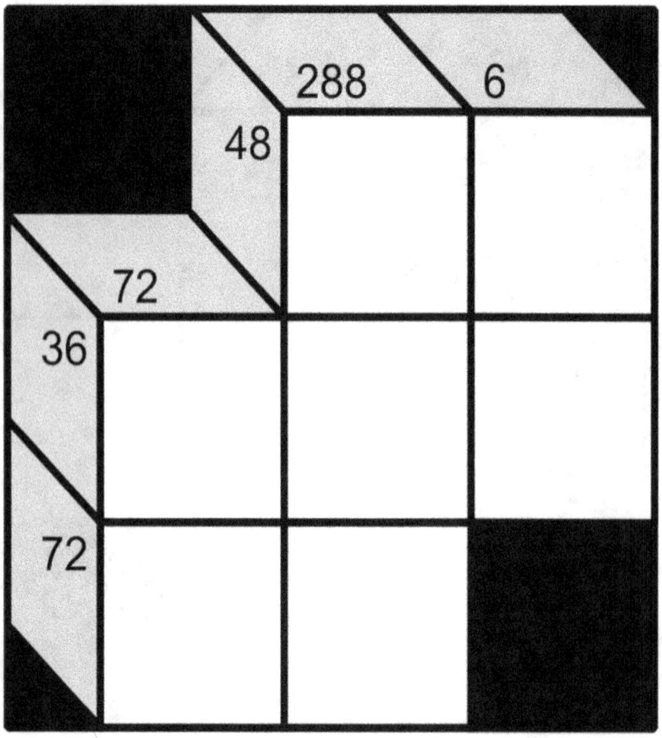

Puzzle 31

Puzzle 32

Puzzle 33

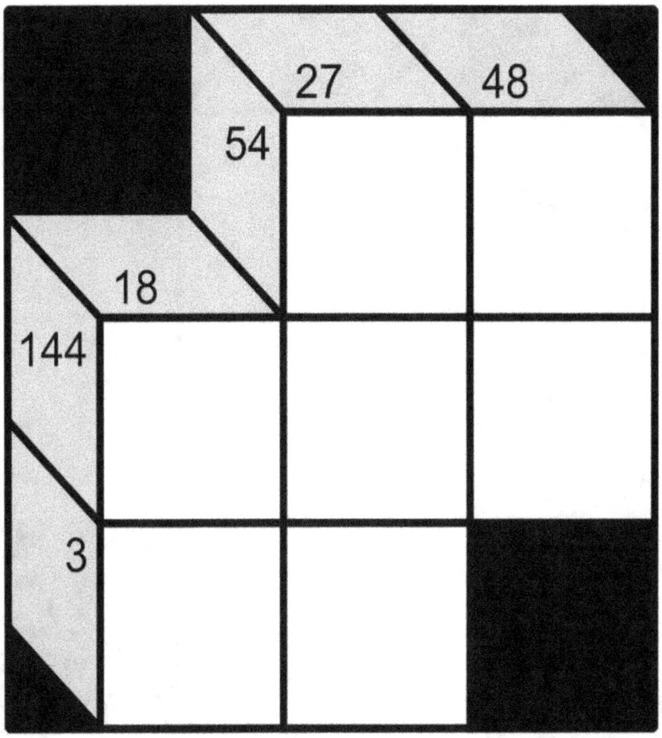

	27	48
54 / 18		
144		
3		■

Puzzle 34

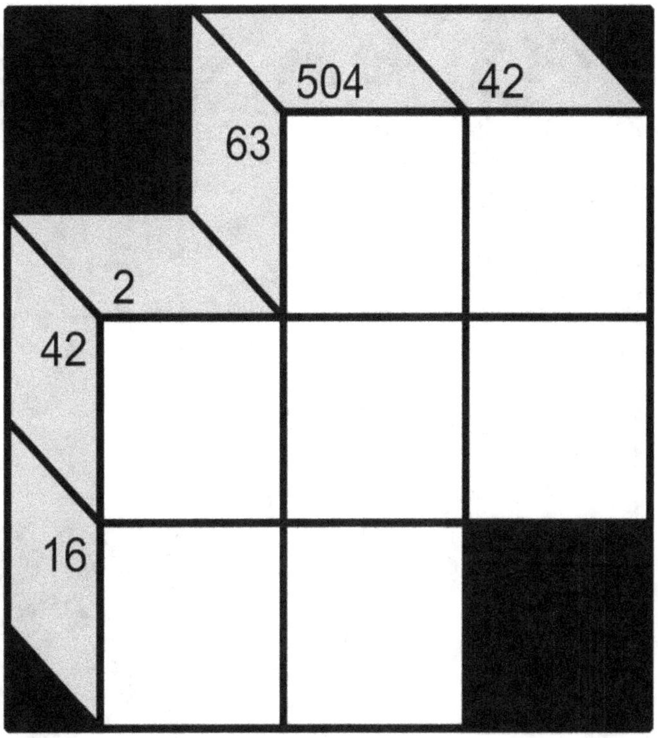

	504	42
63 / 2		
42		
16		■

Puzzle 35

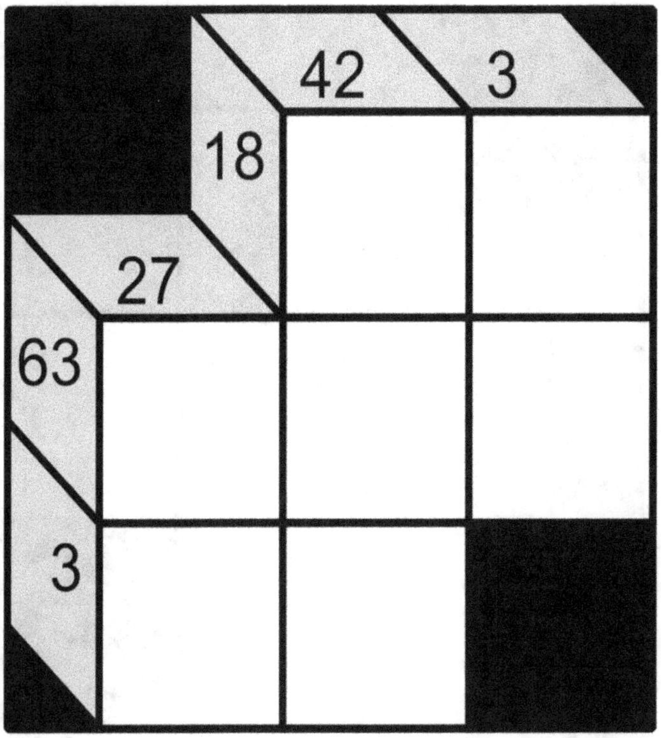

	42	3
18 / 27		
63		
3		■

Puzzle 36

	80	8
10 / 4		
8		
32		■

Puzzle 37

	63	16
56		
72		
18		
72		

Puzzle 38

	432	5
40		
14		
63		
12		

Puzzle 39

	16	28
4		
56		
112		
56		

Puzzle 40

	270	32
40		
27		
216		
27		

Puzzle 41

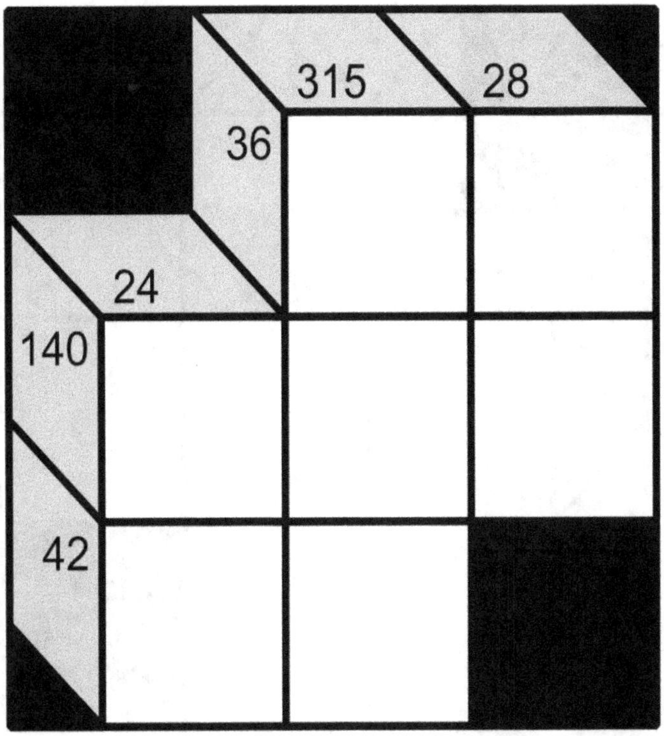

Puzzle 42

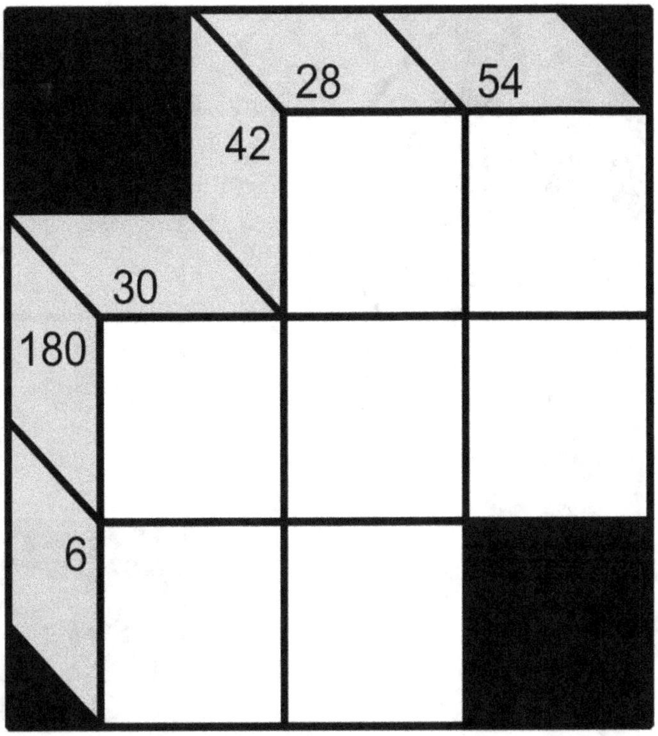

Puzzle 43

Puzzle 44

Puzzle 45

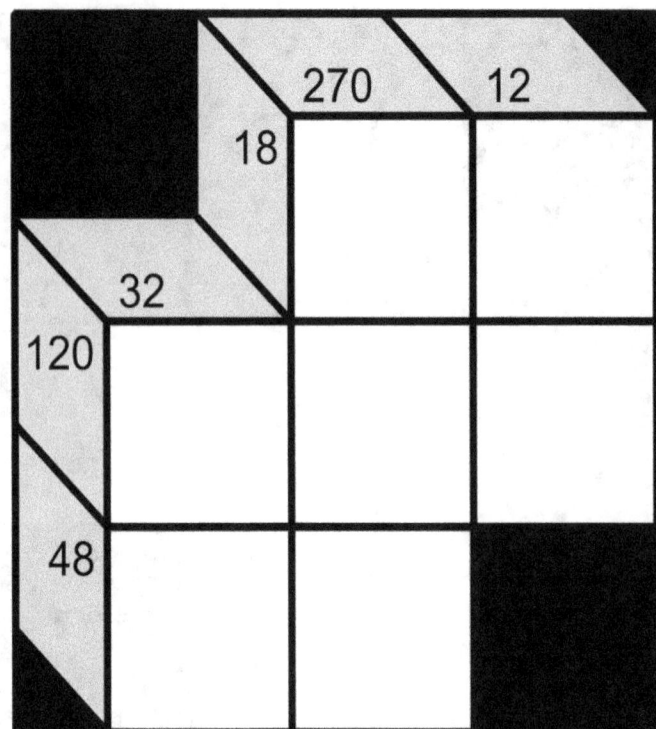

Puzzle 46

Puzzle 47

Puzzle 48

Puzzle 49

Puzzle 50

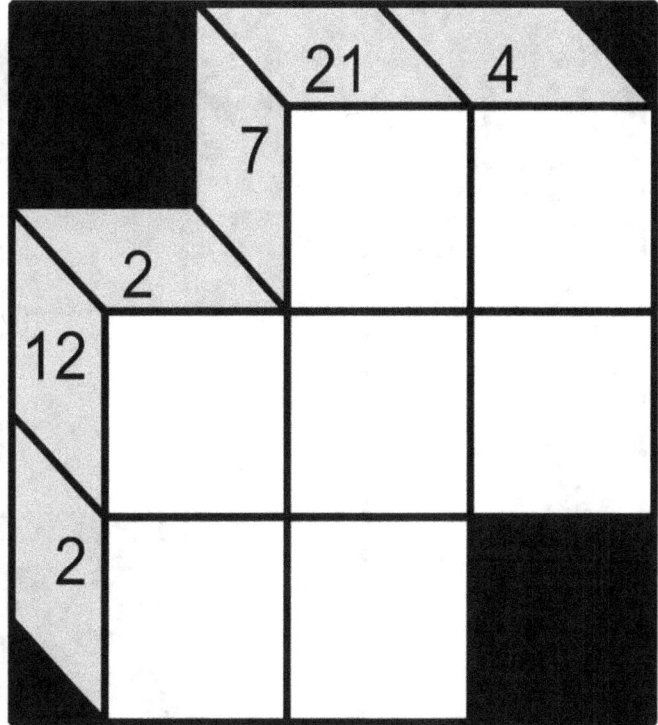

Puzzle 51

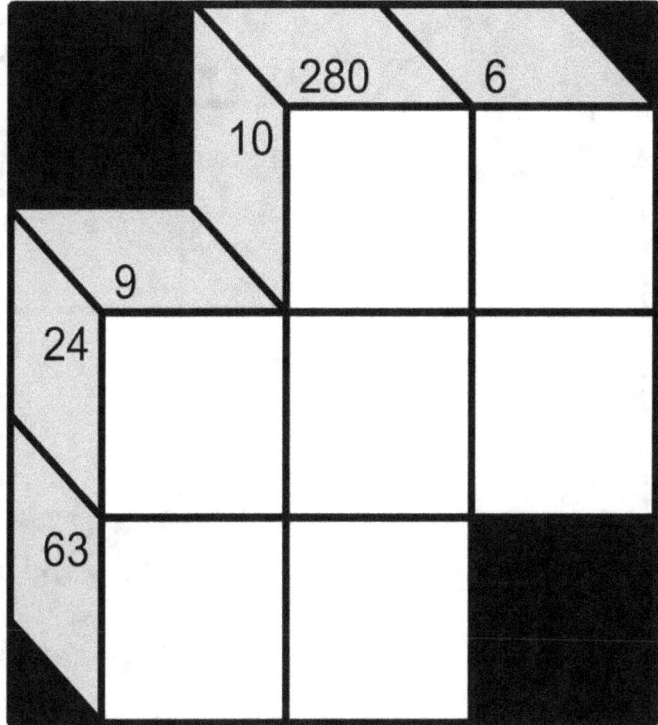

Puzzle 52

Puzzle 53

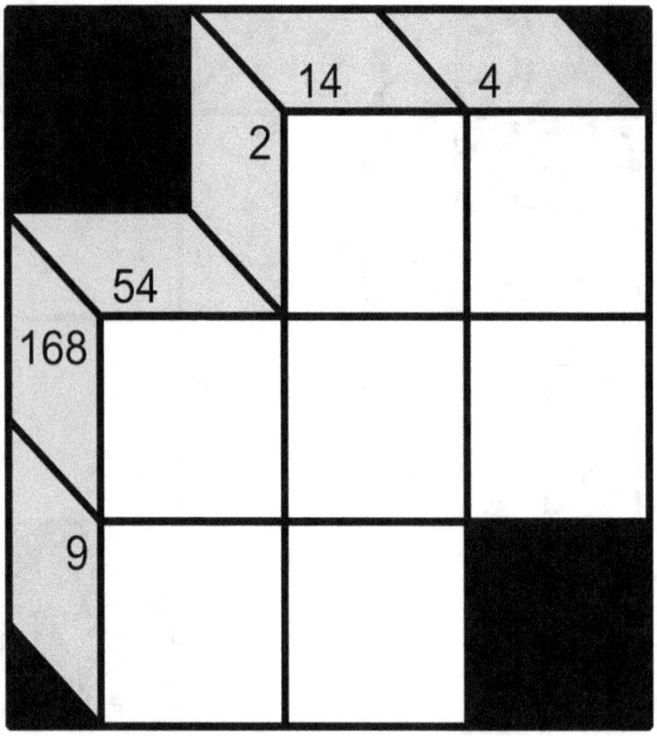

Puzzle 54

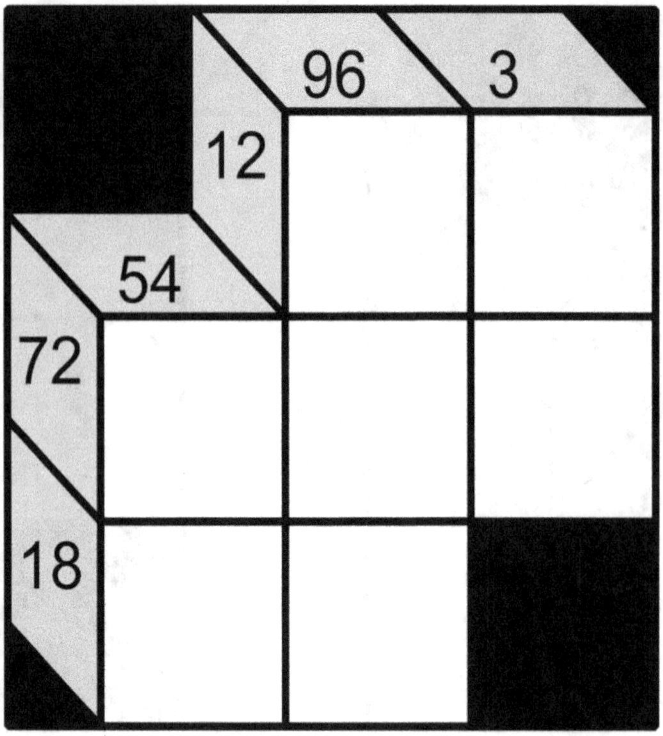

Puzzle 55

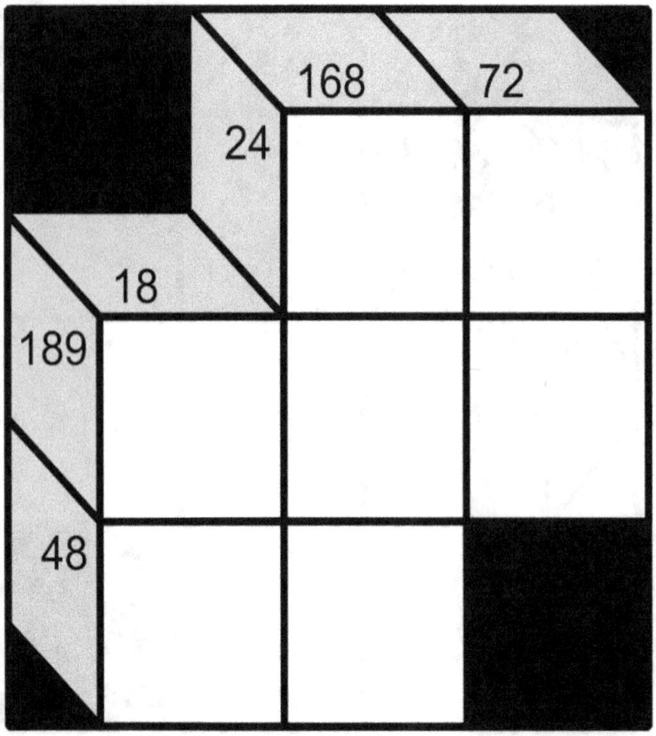

Puzzle 56

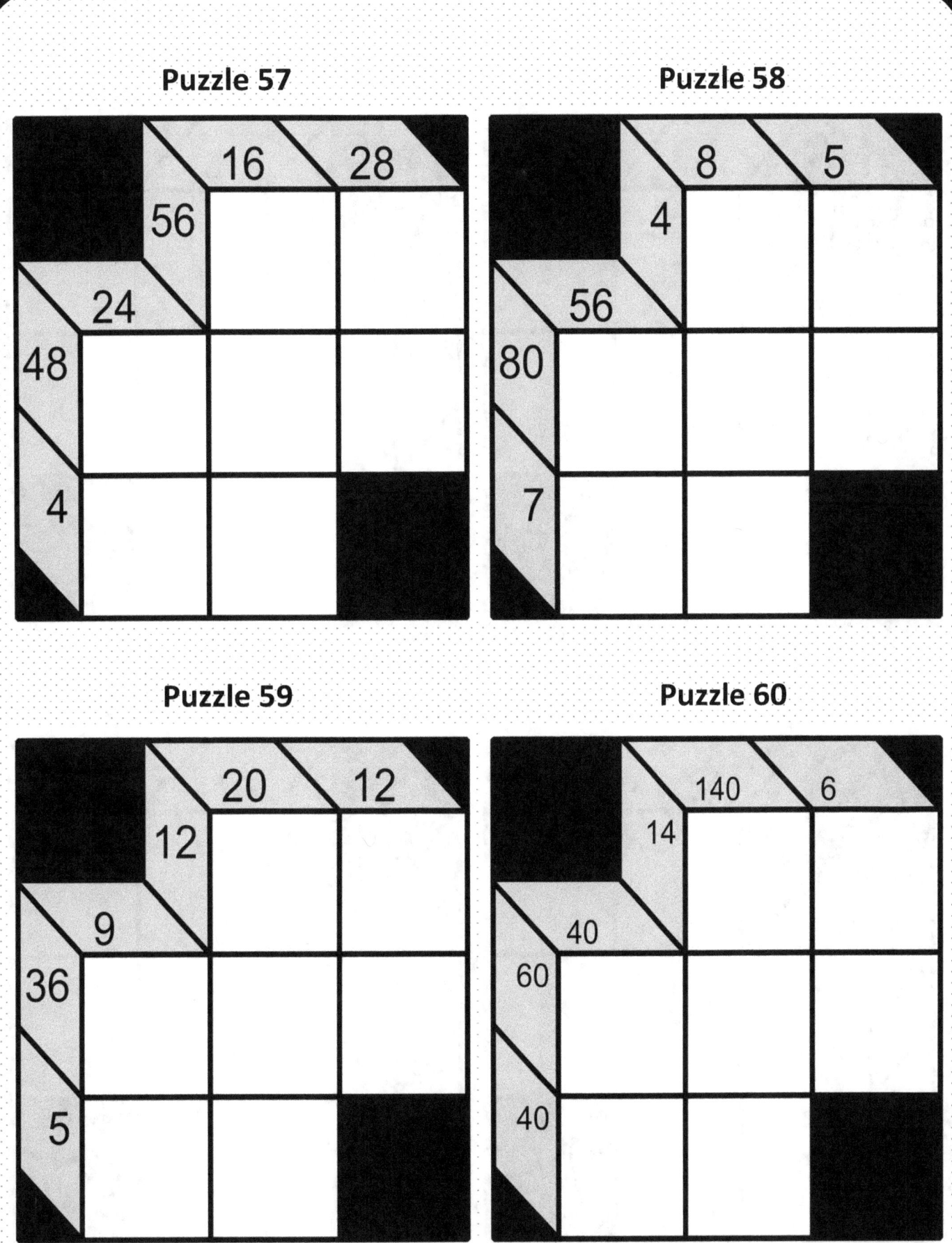

Puzzle 61

Puzzle 62

Puzzle 63

Puzzle 64

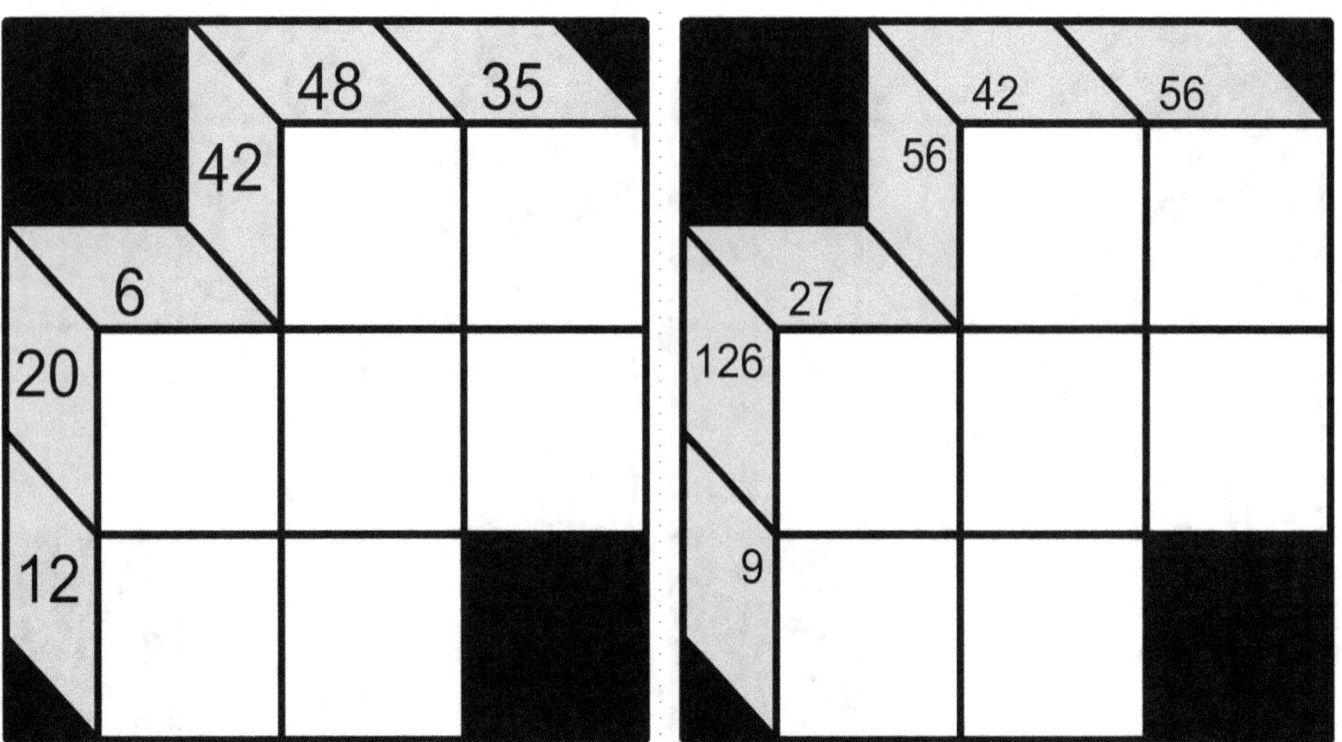

Puzzle 65

Puzzle 66

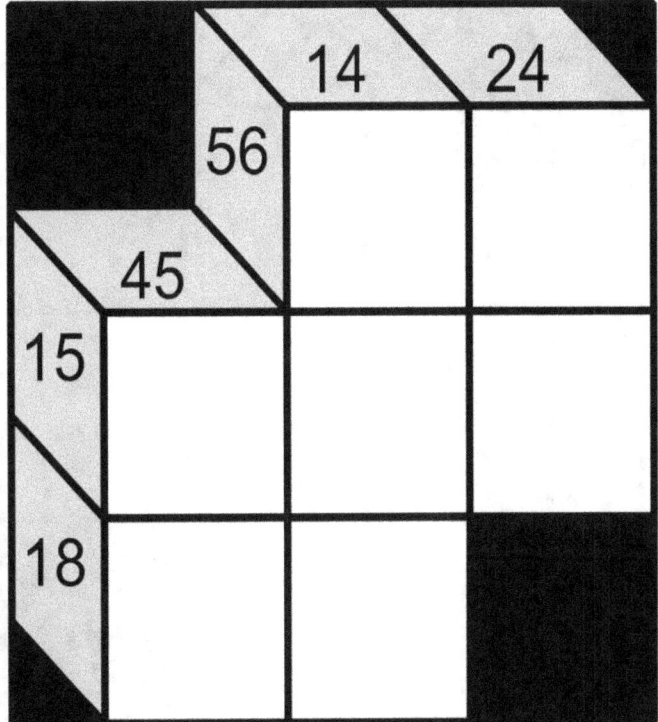

Puzzle 67

Puzzle 68

Puzzle 69

Puzzle 70

Puzzle 71

Puzzle 72

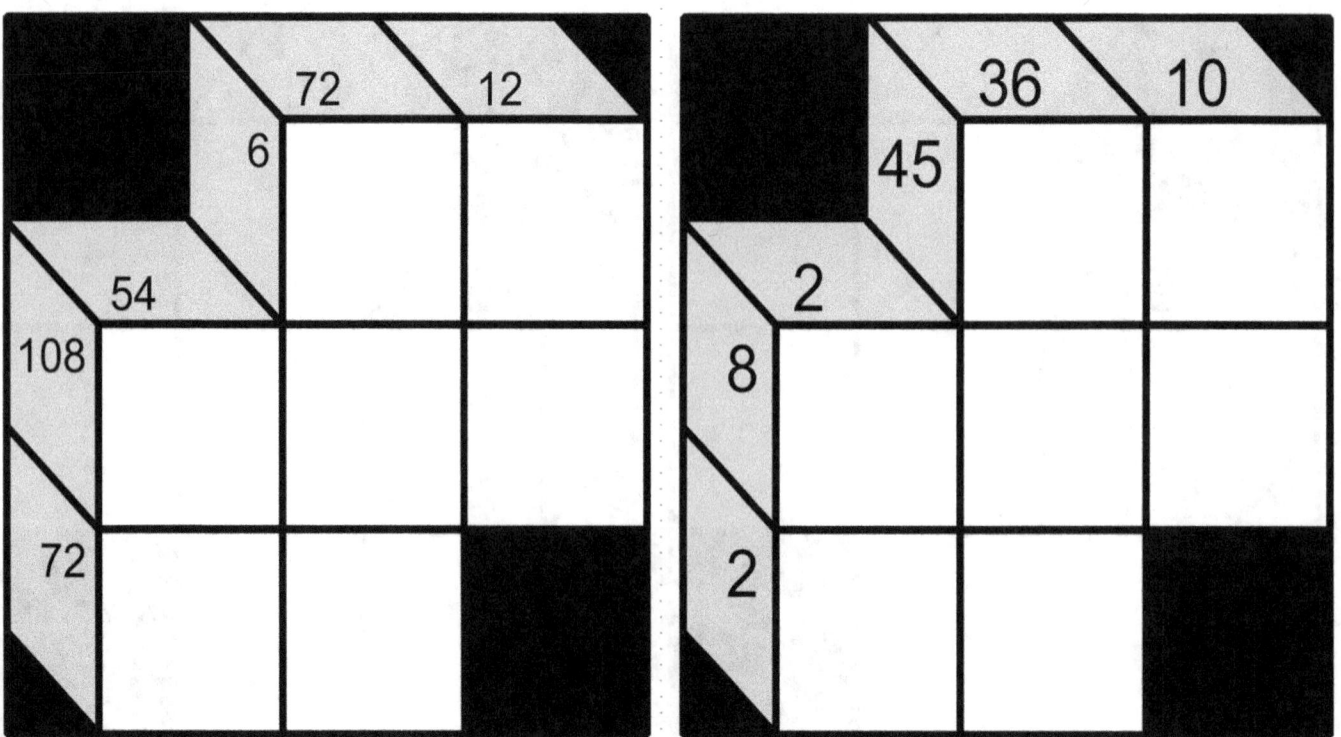

Puzzle 73

Puzzle 74

Puzzle 75

Puzzle 76

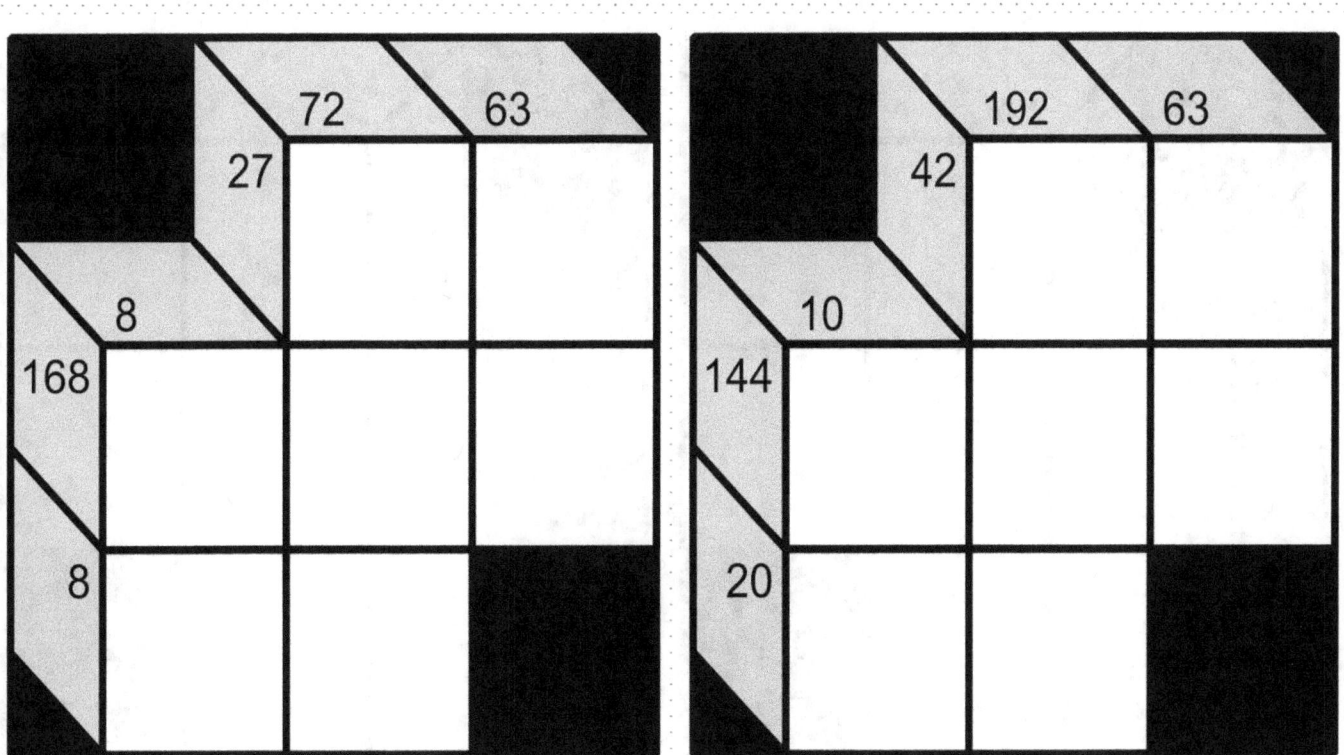

Puzzle 77

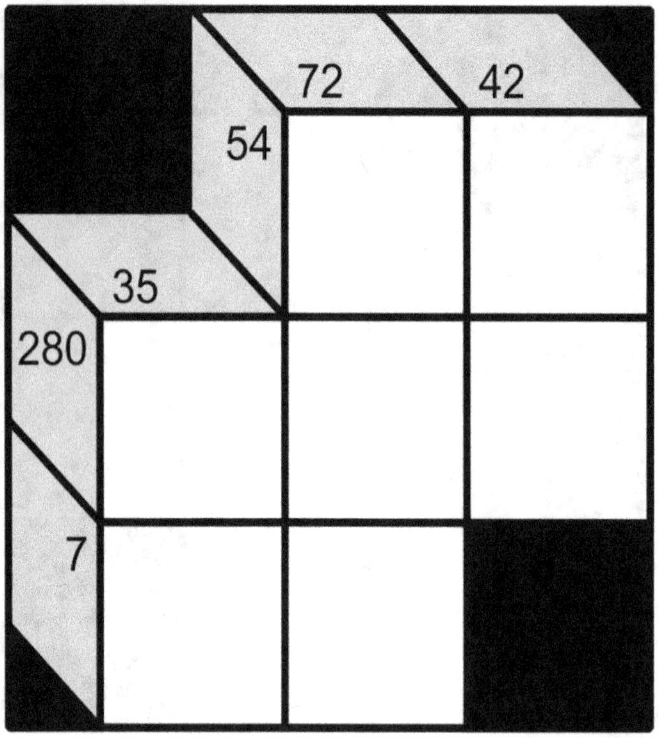

Puzzle 78

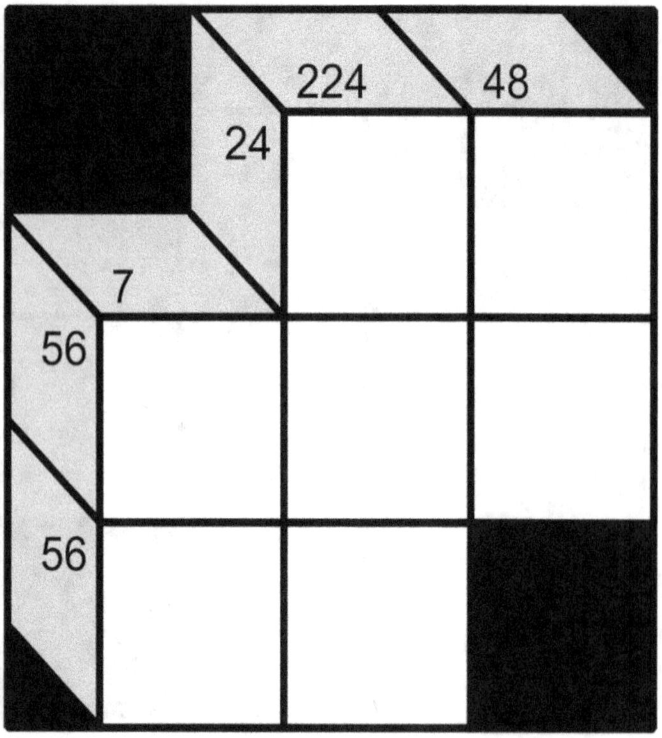

Puzzle 79

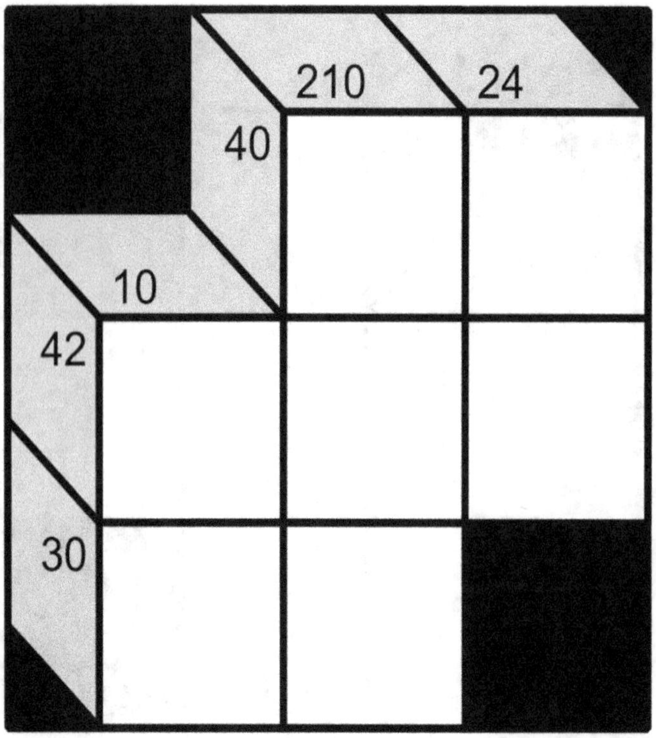

Puzzle 80

Puzzle 81

Puzzle 82

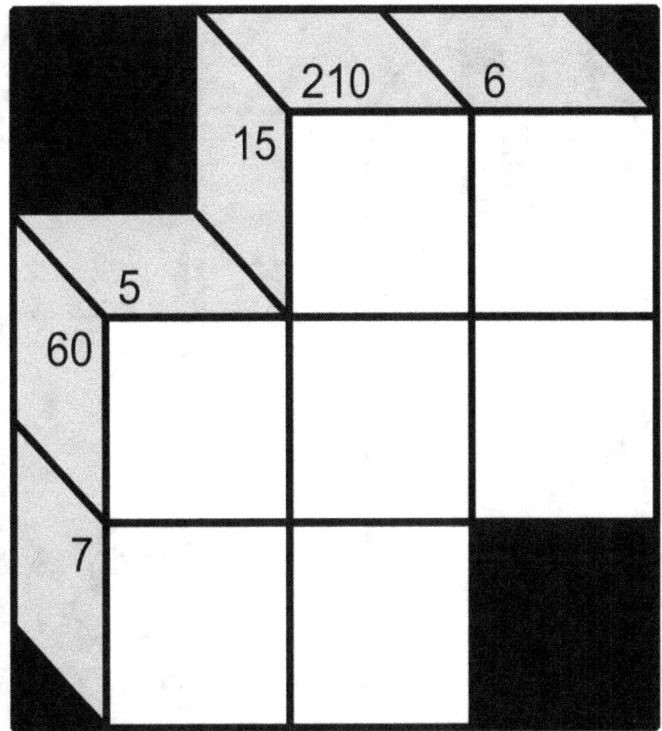

Puzzle 83

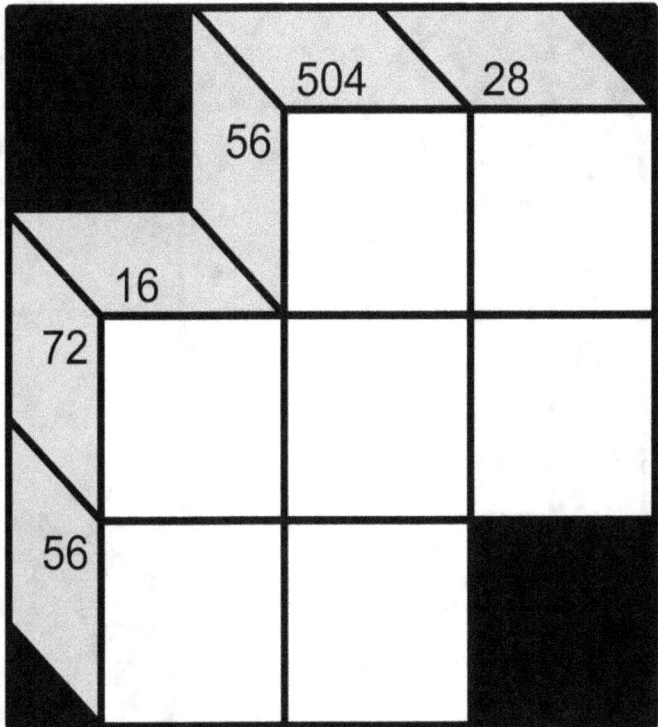

Puzzle 84

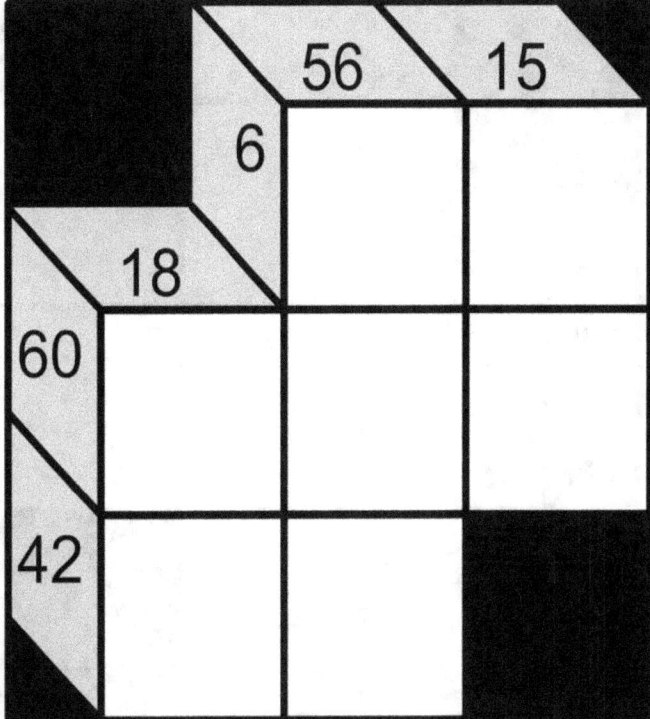

Puzzle 85

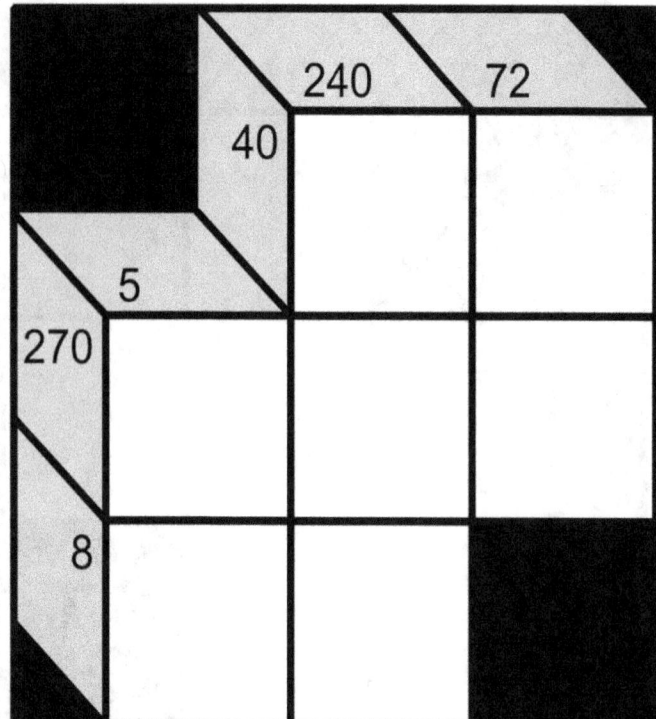

Puzzle 86

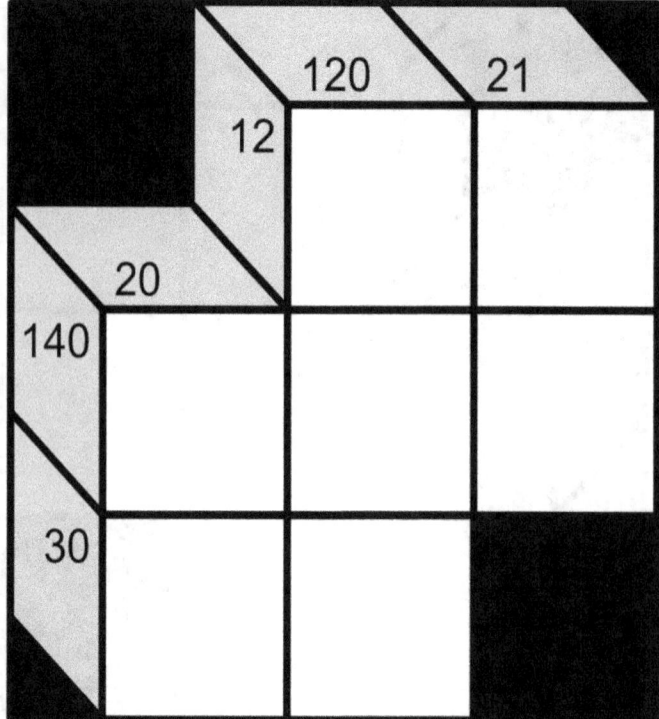

Puzzle 87

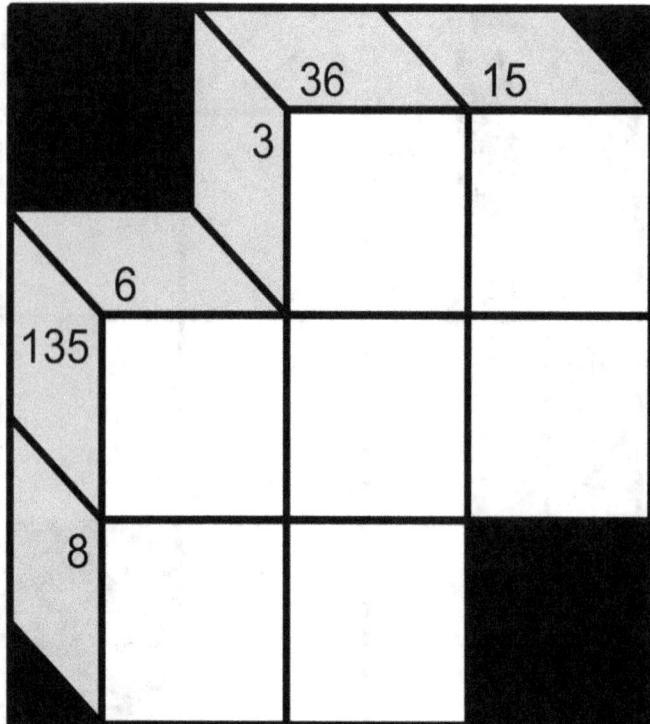

Puzzle 88

Puzzle 89

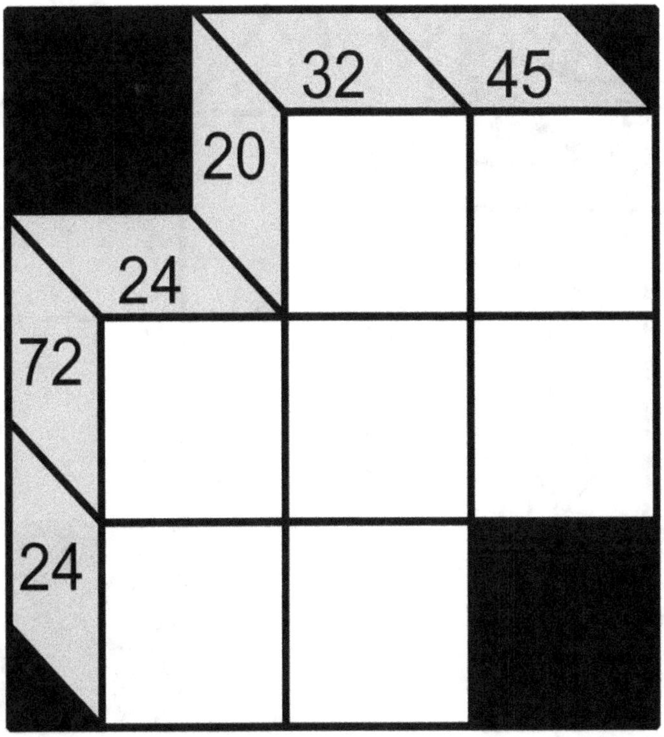

Puzzle 90

Puzzle 91

Puzzle 92

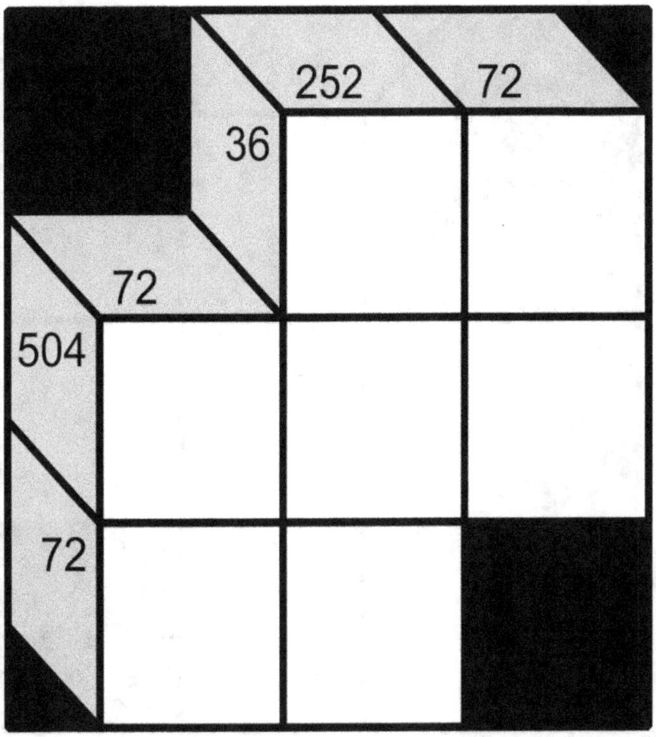

Puzzle 93

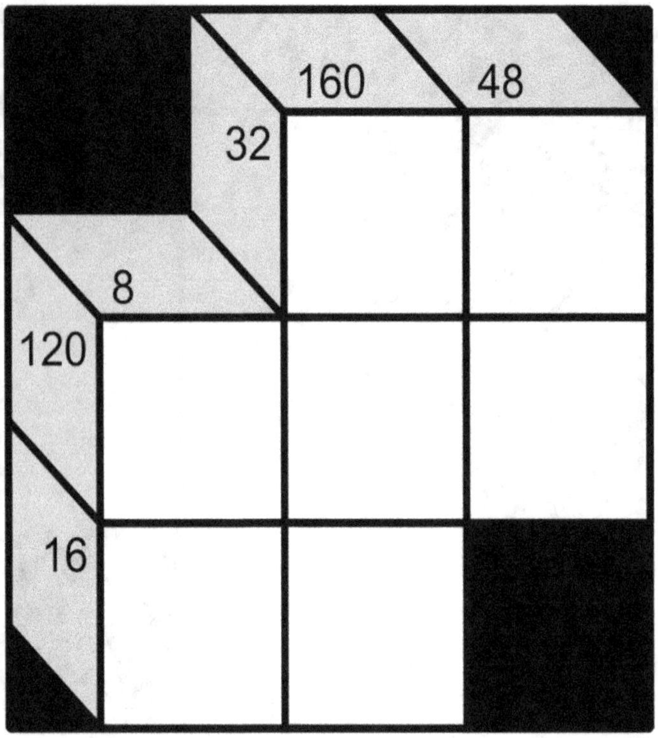

Puzzle 94

Puzzle 95

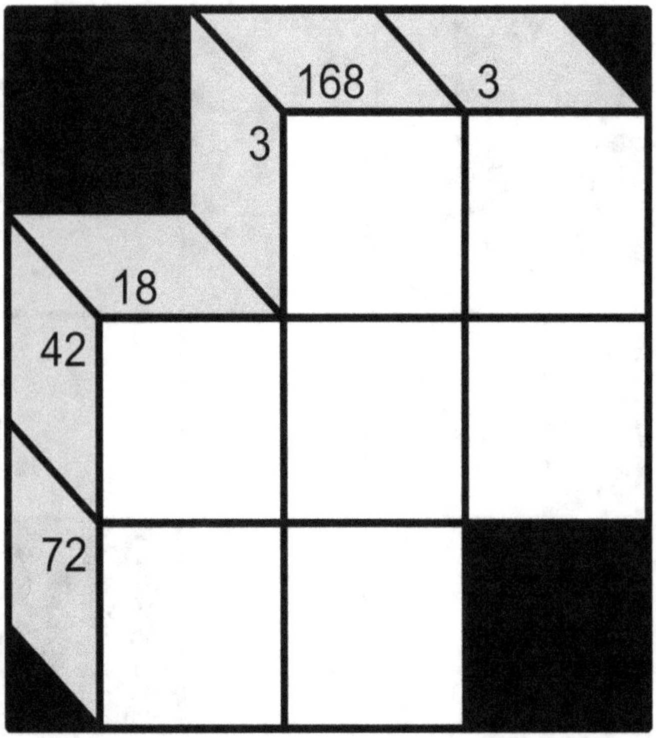

Puzzle 96

Puzzle 97

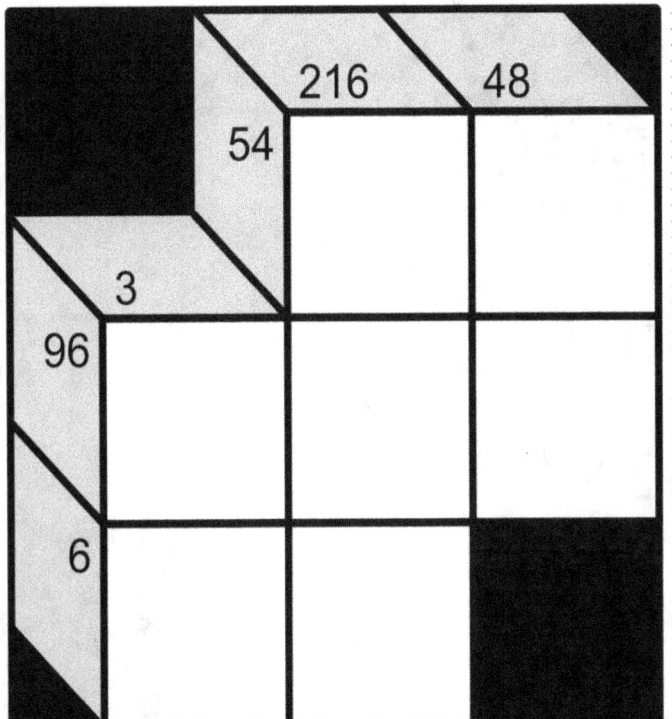

Puzzle 98

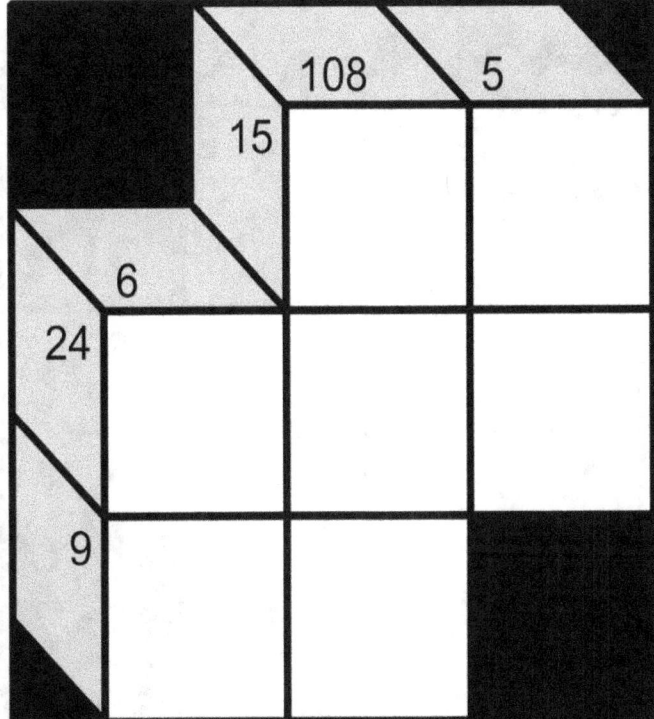

Puzzle 99

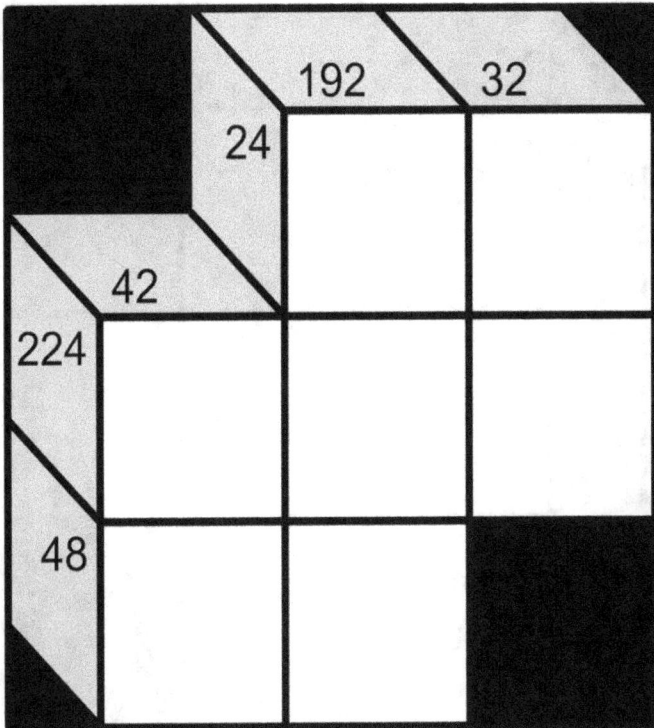

Puzzle 100

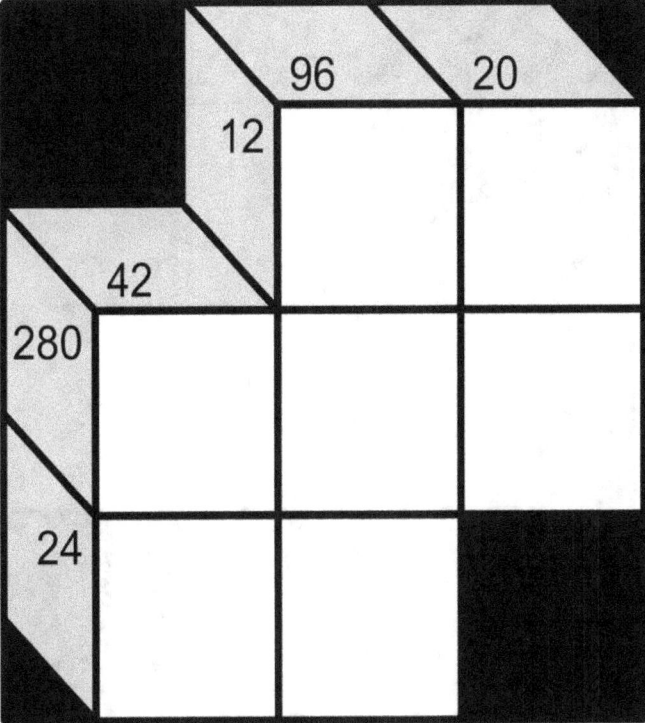

Puzzle 101
Puzzle 102
Puzzle 103
Puzzle 104

Puzzle 105

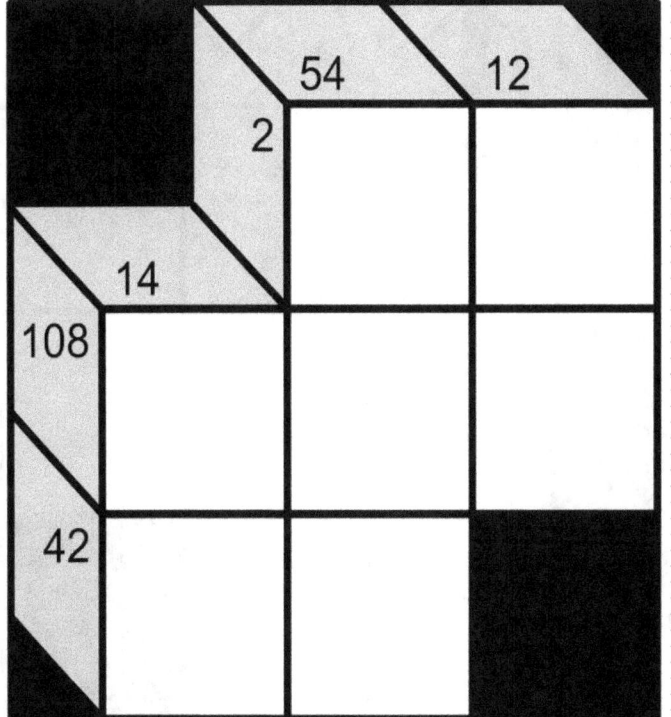

Puzzle 106

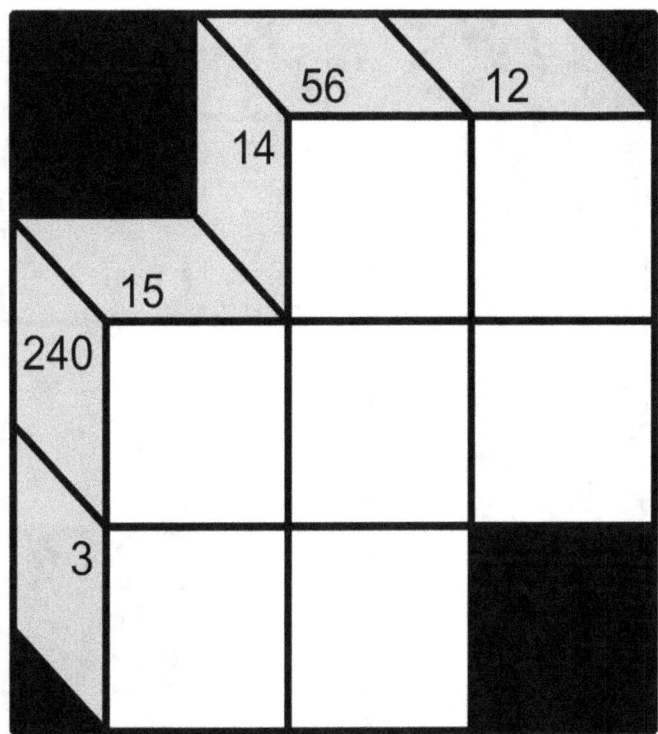

Puzzle 107

Puzzle 108

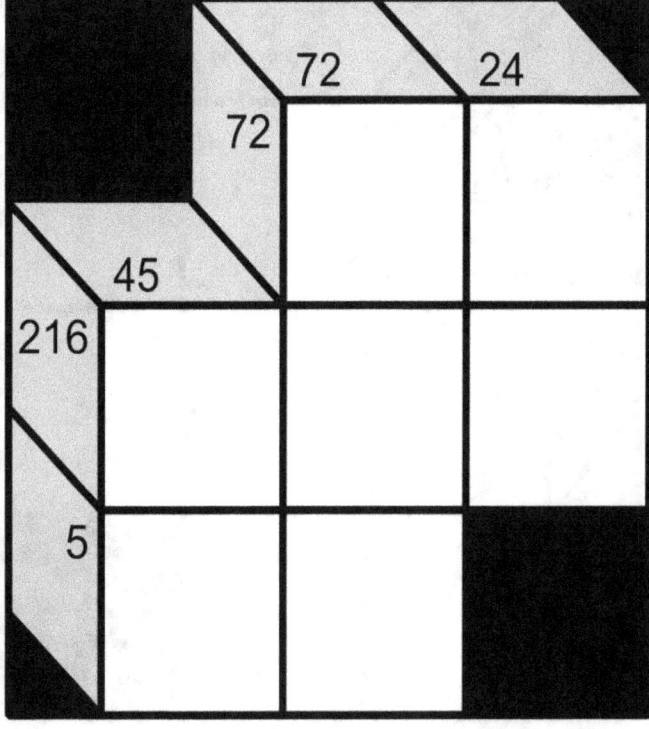

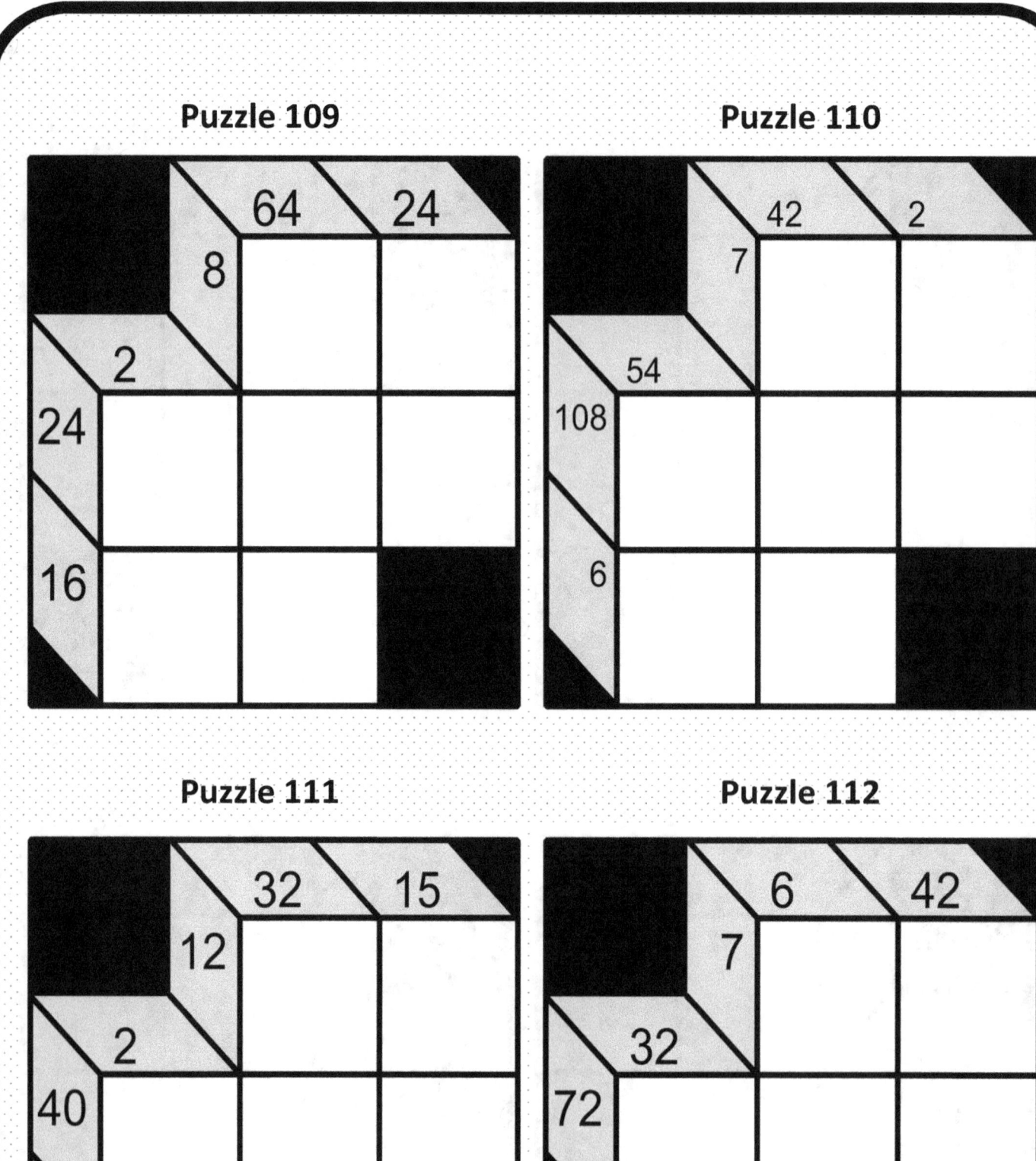

Puzzle 113

Puzzle 114

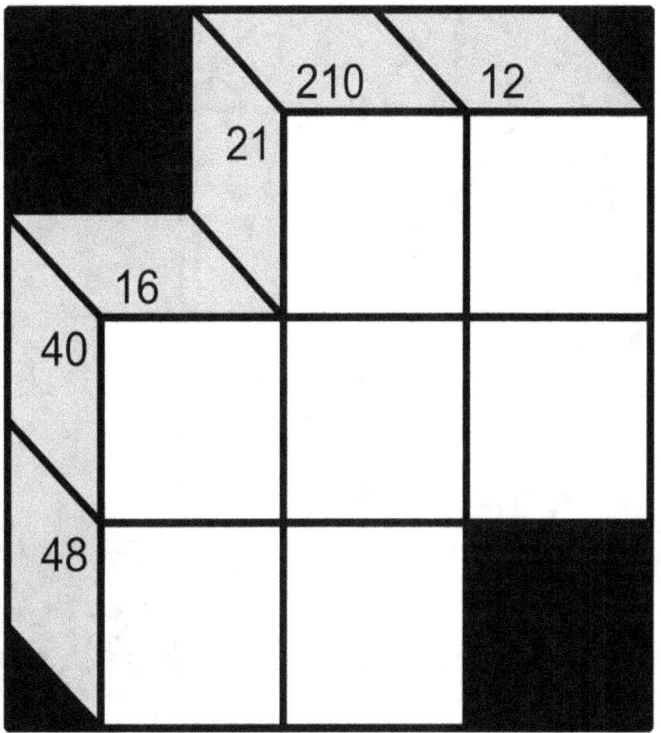

Puzzle 115

Puzzle 116

Puzzle 117

Puzzle 118

Puzzle 119

Puzzle 120

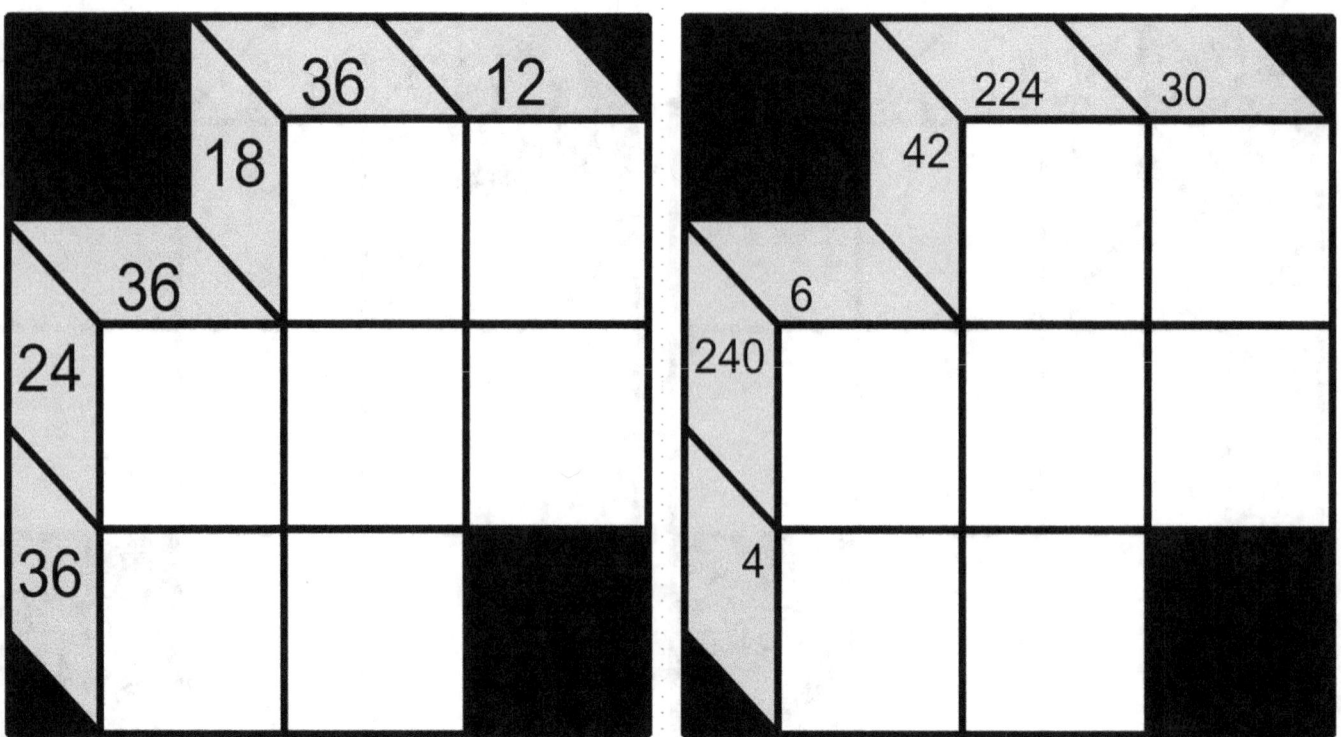

Puzzle 121

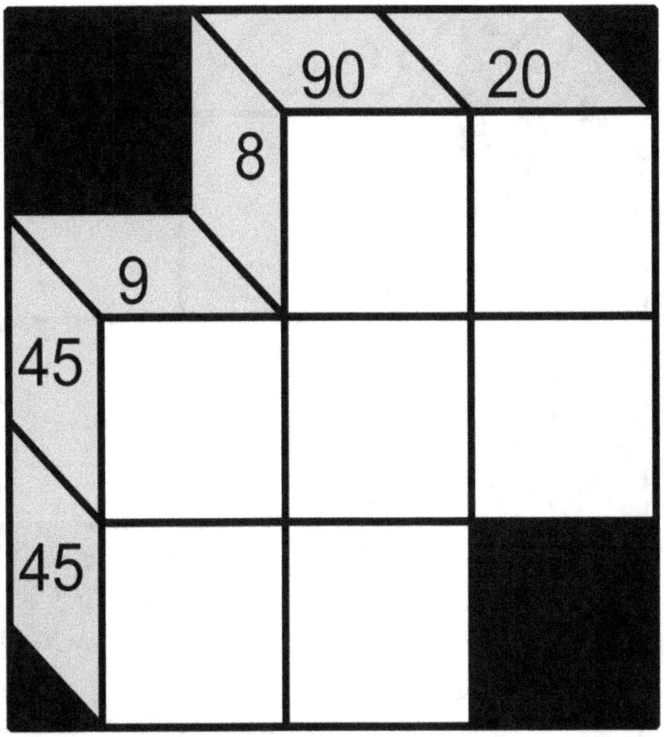

Puzzle 122

Puzzle 123

Puzzle 124

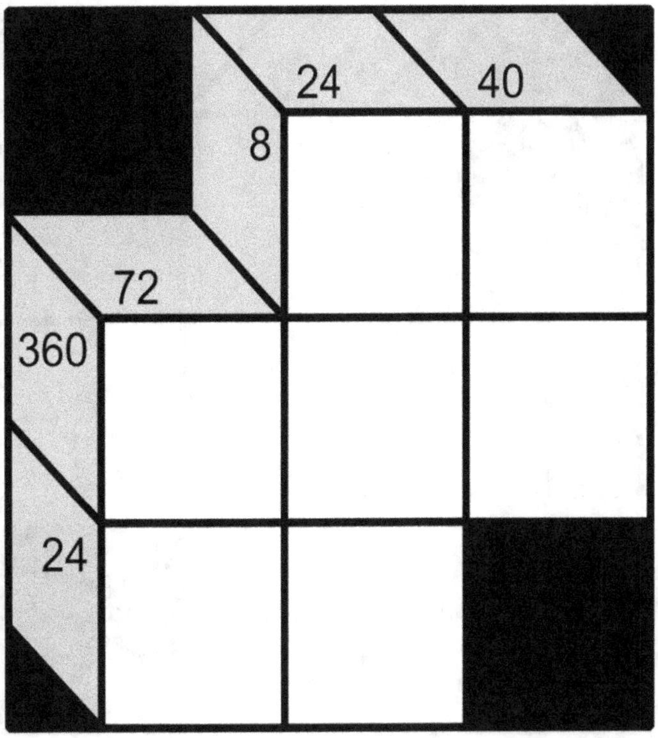

Puzzle 125

Puzzle 126

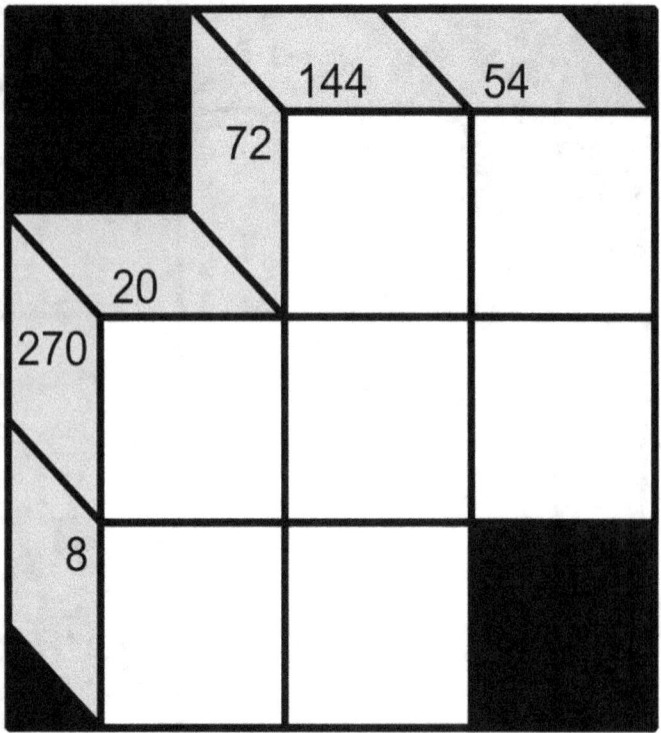

Puzzle 127

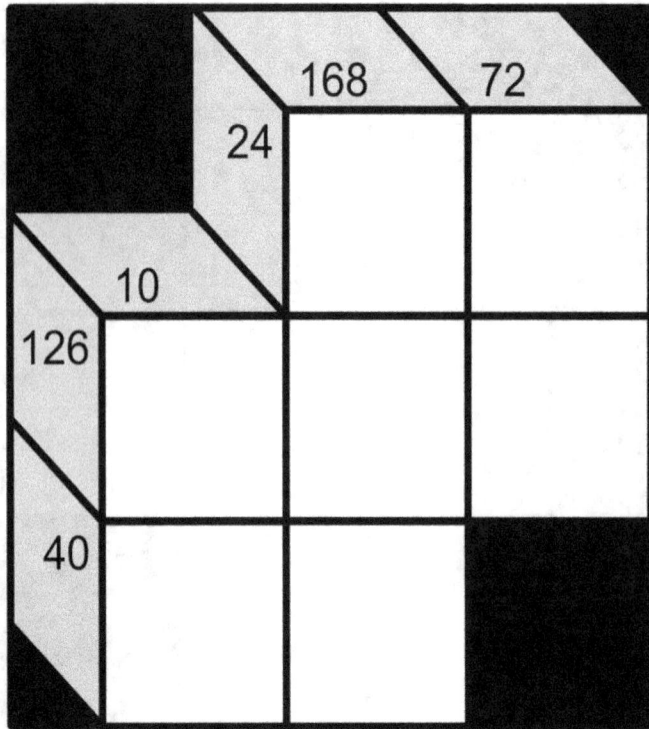

Puzzle 128

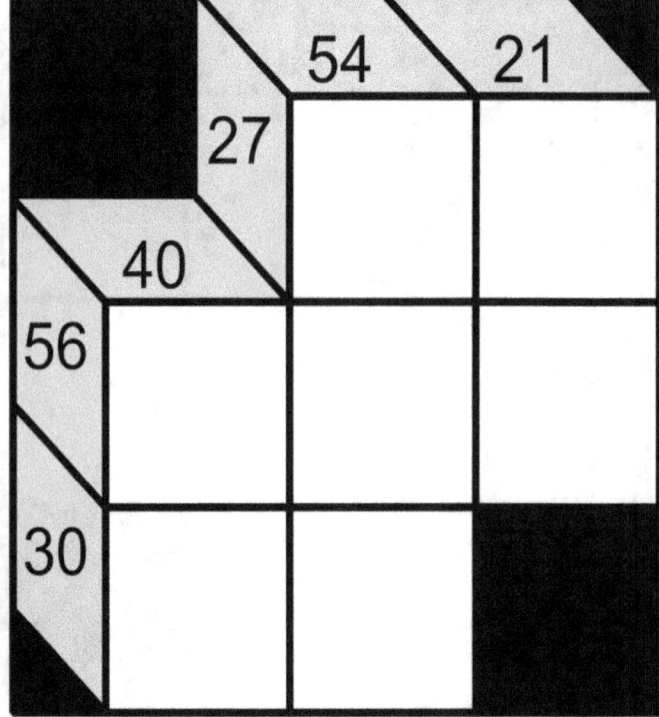

Puzzle 129

Puzzle 130

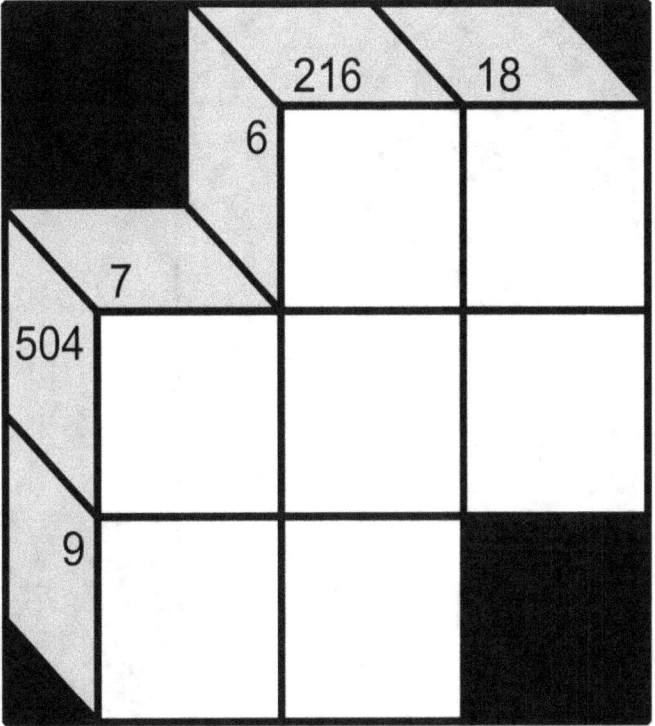

Puzzle 131

Puzzle 132

Puzzle 133

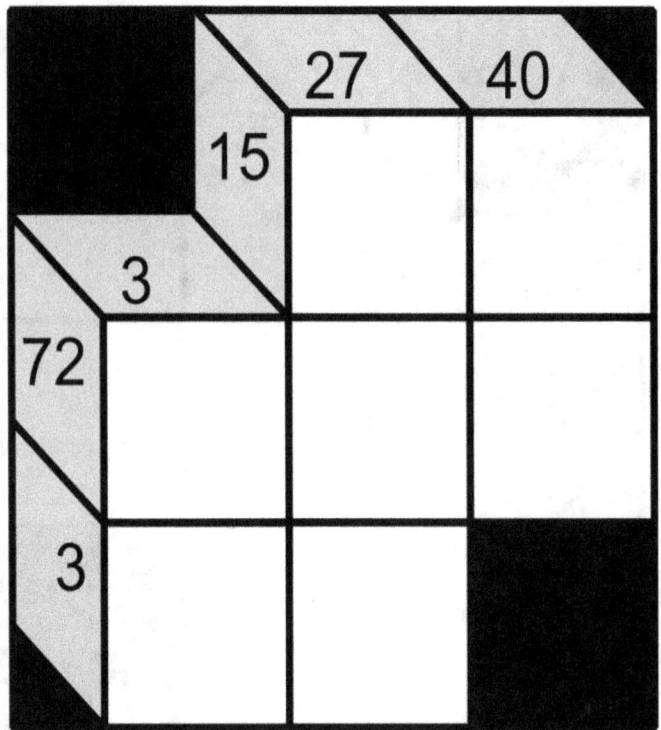

Puzzle 134

Puzzle 135

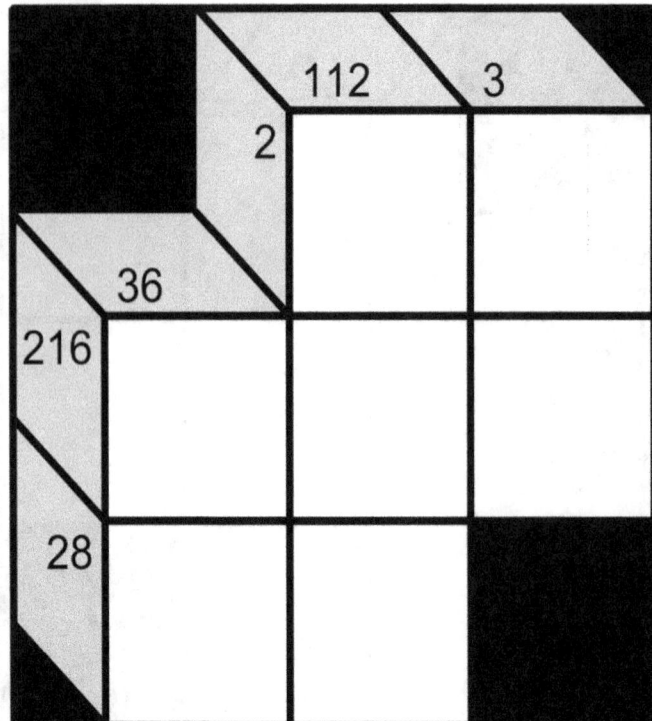

Puzzle 136

Puzzle 137

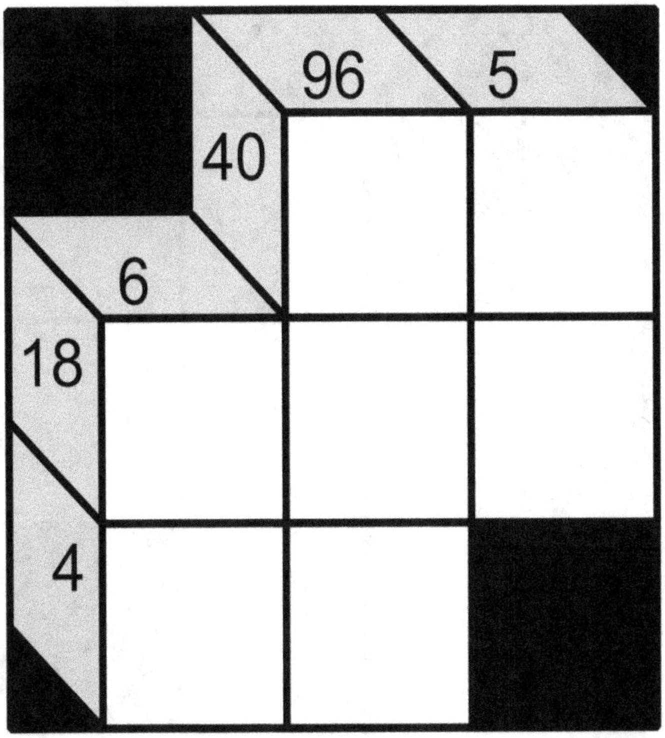

Puzzle 138

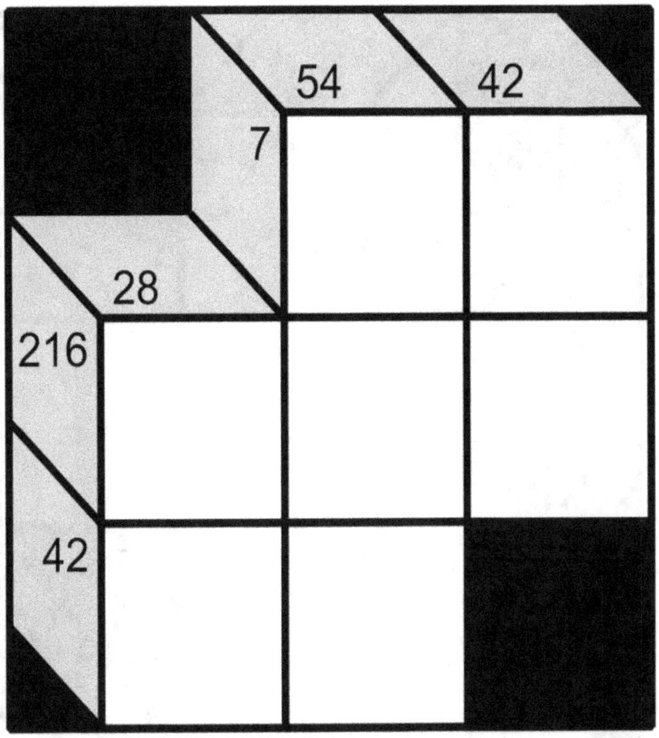

Puzzle 139

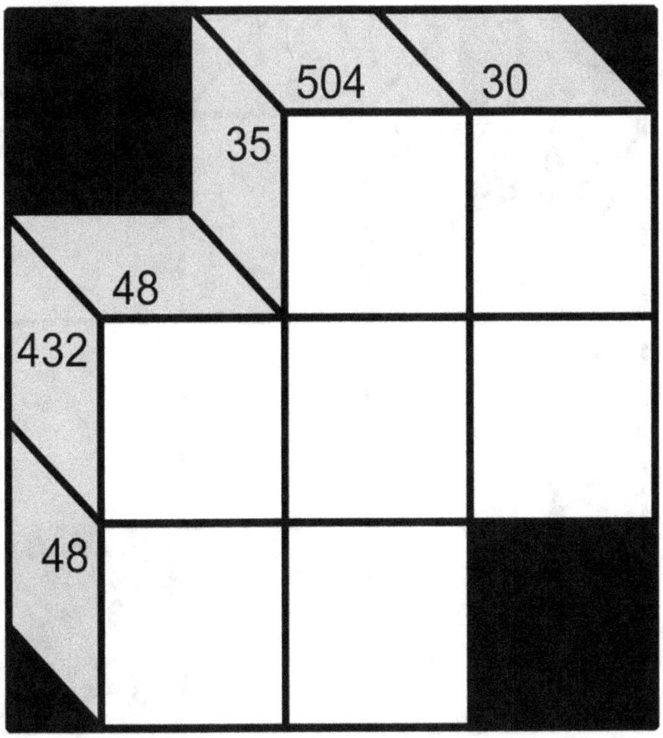

Puzzle 140

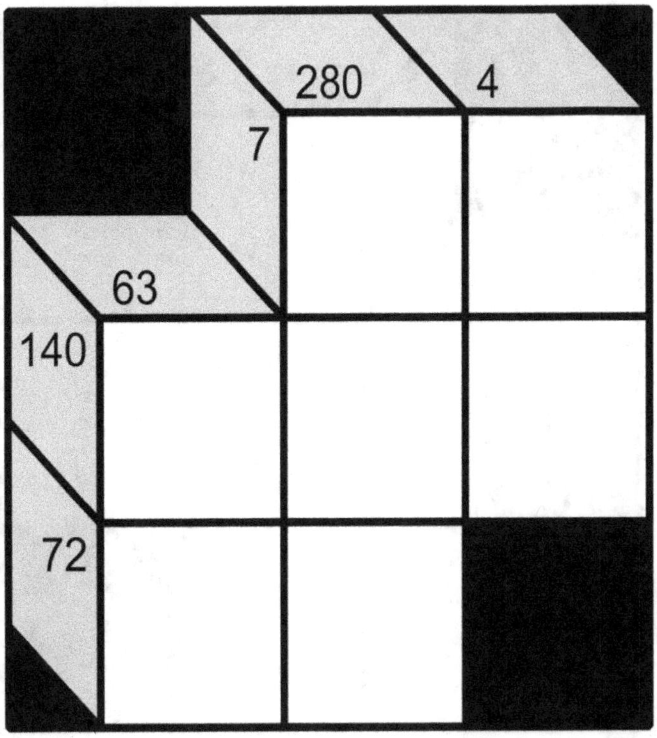

Puzzle 141

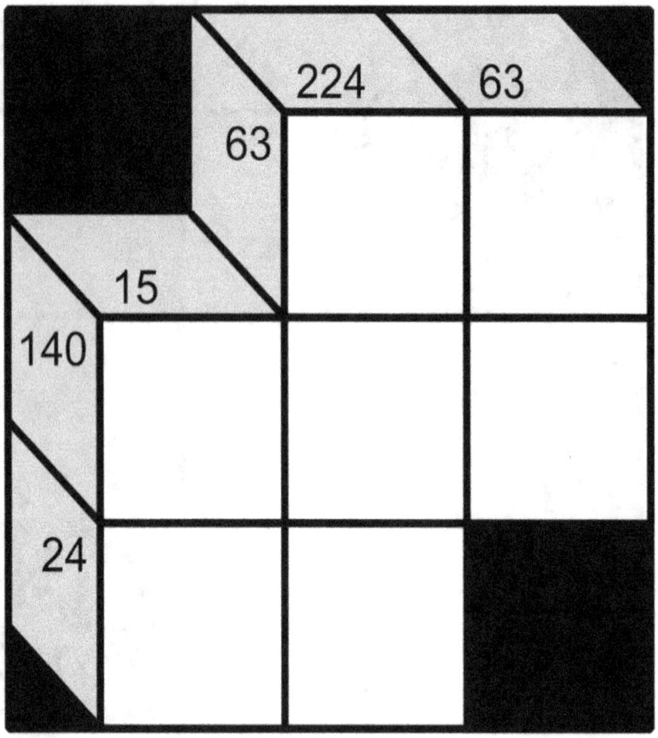
224
63
63
15
140
24

Puzzle 142

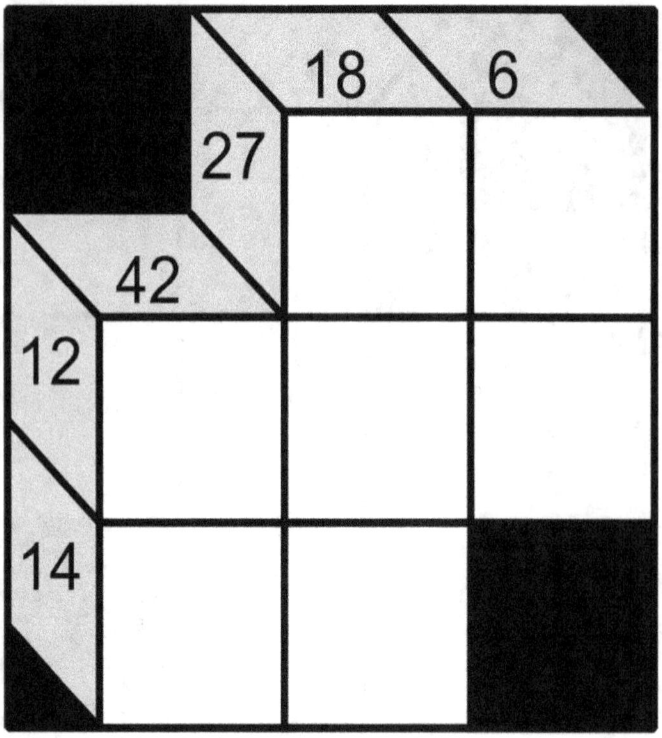
18
6
27
42
12
14

Puzzle 143

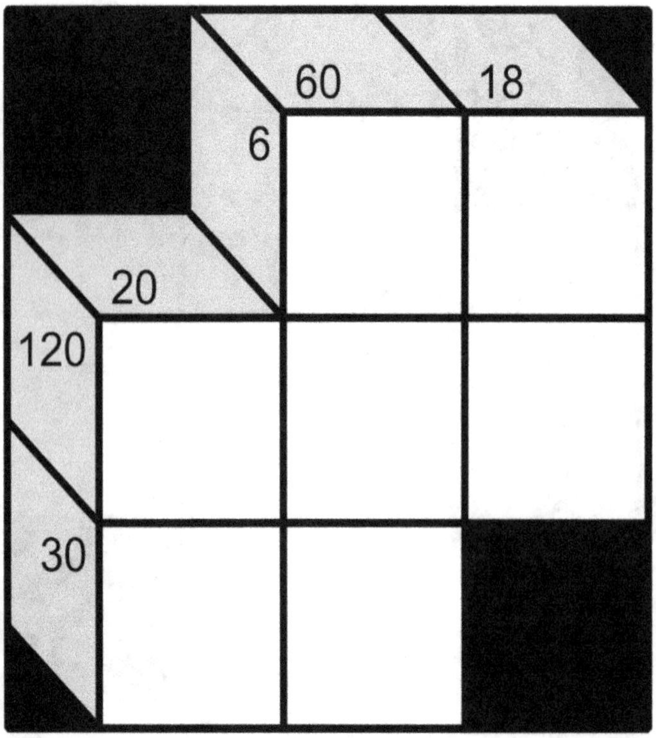
60
18
6
20
120
30

Puzzle 144

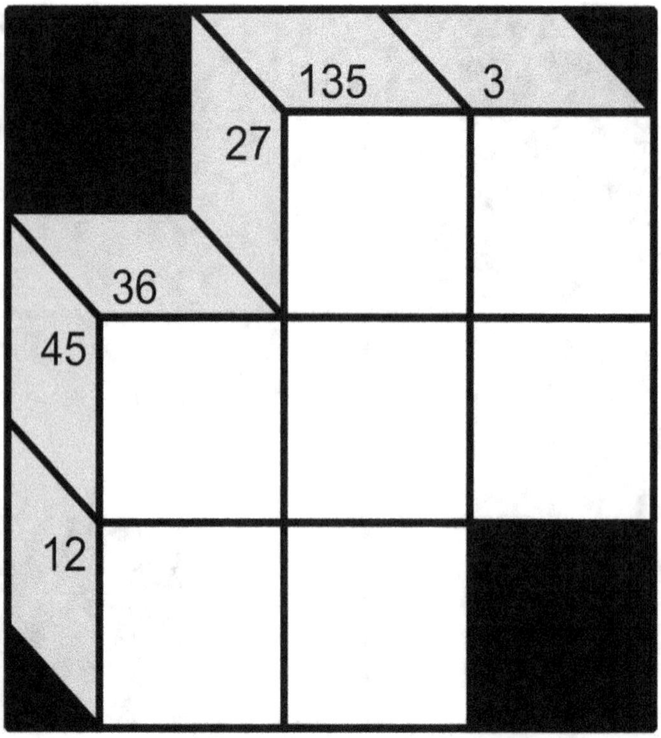
135
3
27
36
45
12

Puzzle 145

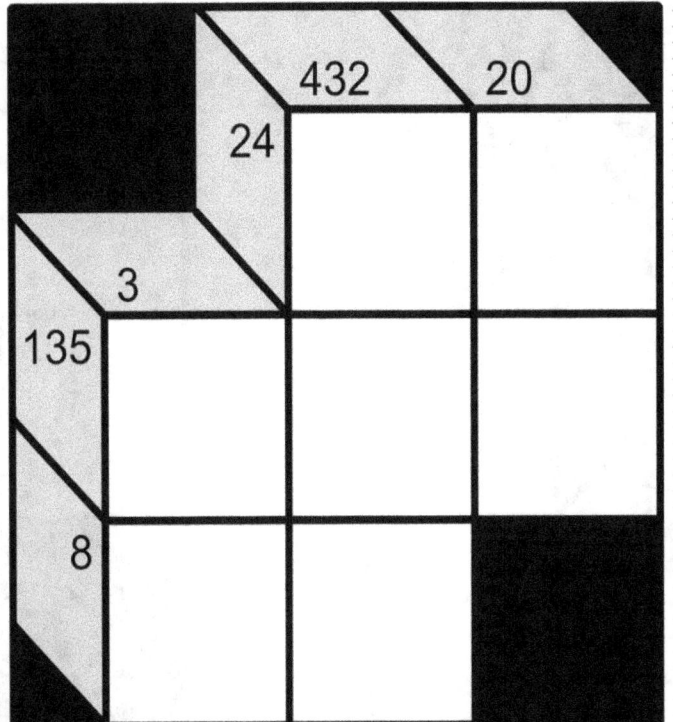

Puzzle 146

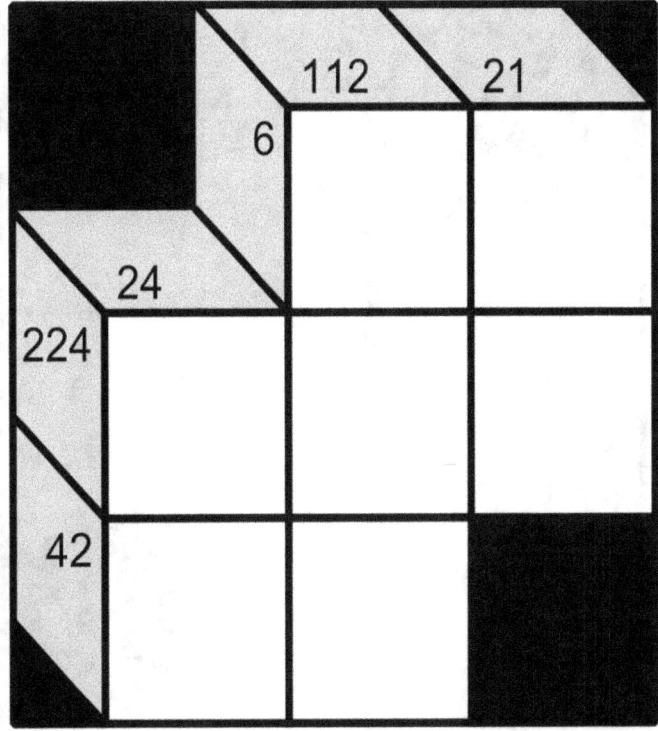

Puzzle 147

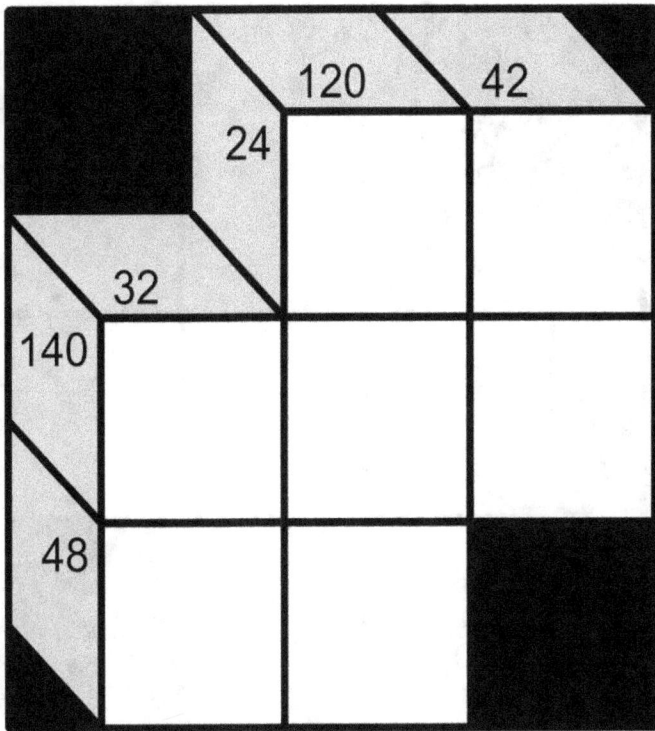

Puzzle 148

Puzzle 149 **Puzzle 150**

Puzzle 151 **Puzzle 152**

Puzzle 153

	108	7
3		
6		
126		
12		

Puzzle 154

	96	5
40		
20		
8		
30		

Puzzle 155

	16	30
5		
45		
432		
10		

Puzzle 156

	6	24
4		
40		
60		
24		

Puzzle 157

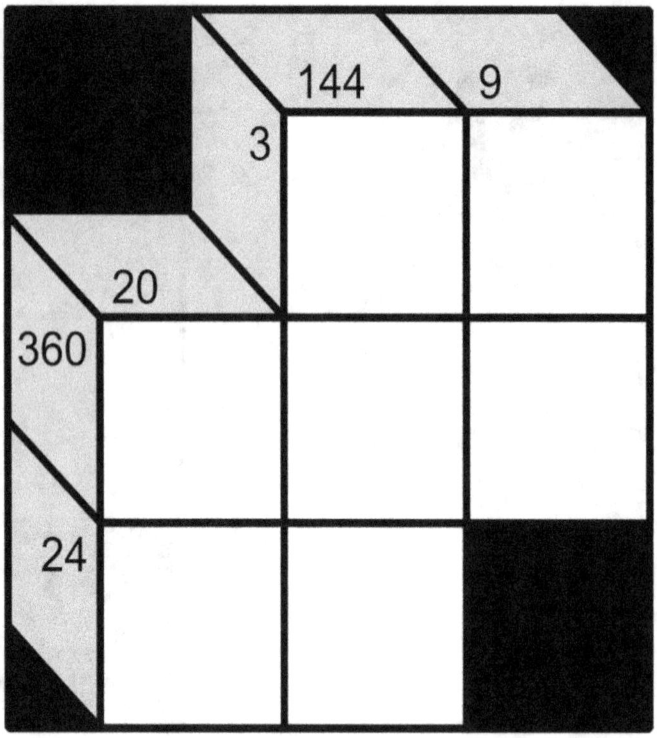
144
9
3
20
360
24

Puzzle 158

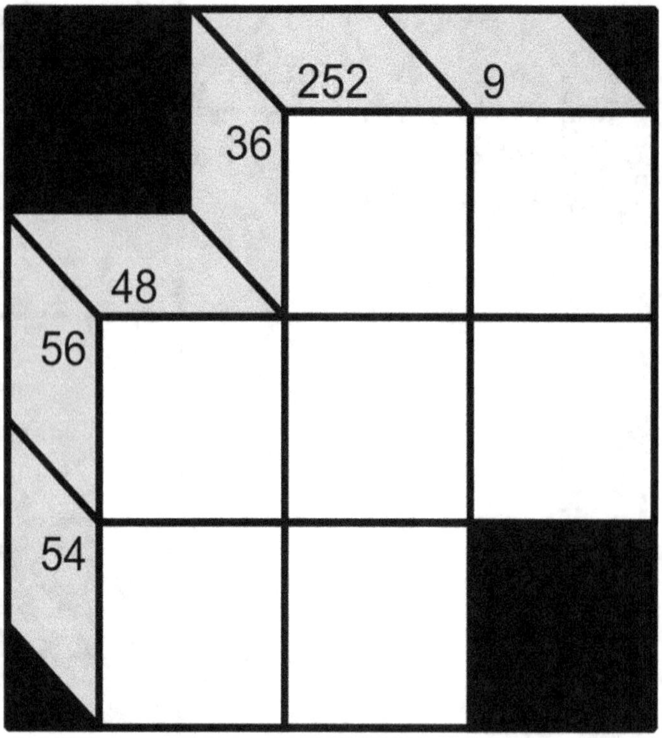
252
9
36
48
56
54

Puzzle 159

84
40
24
6
40
21

Puzzle 160

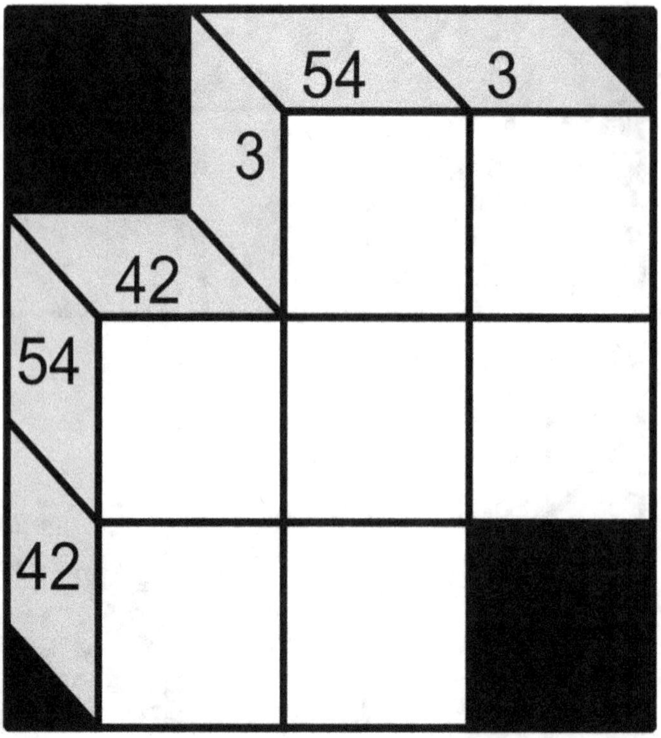
54
3
3
42
54
42

Puzzle 161

Puzzle 162

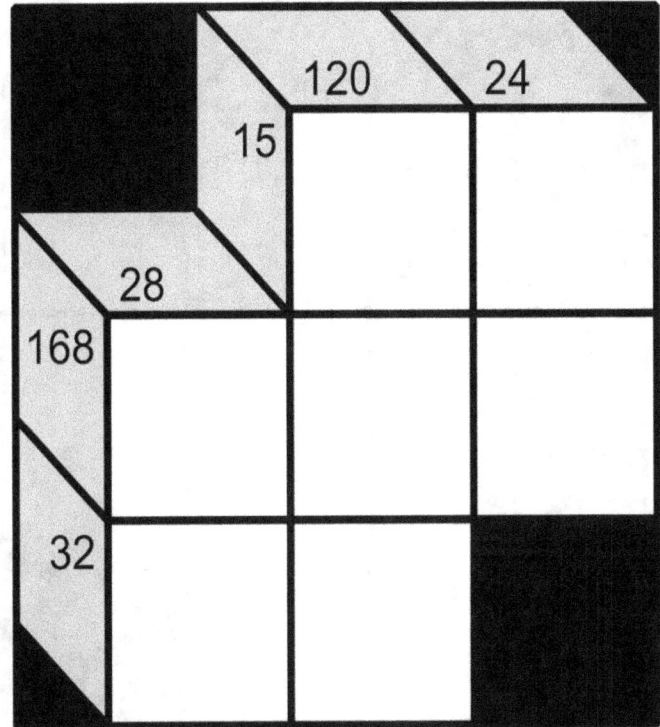

Puzzle 163

Puzzle 164

Puzzle 165

Puzzle 166

Puzzle 167

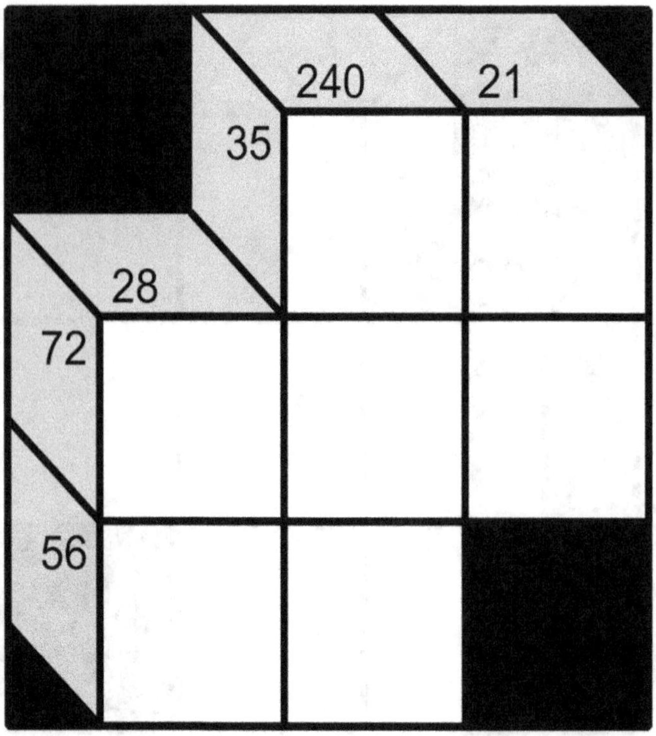

Puzzle 168

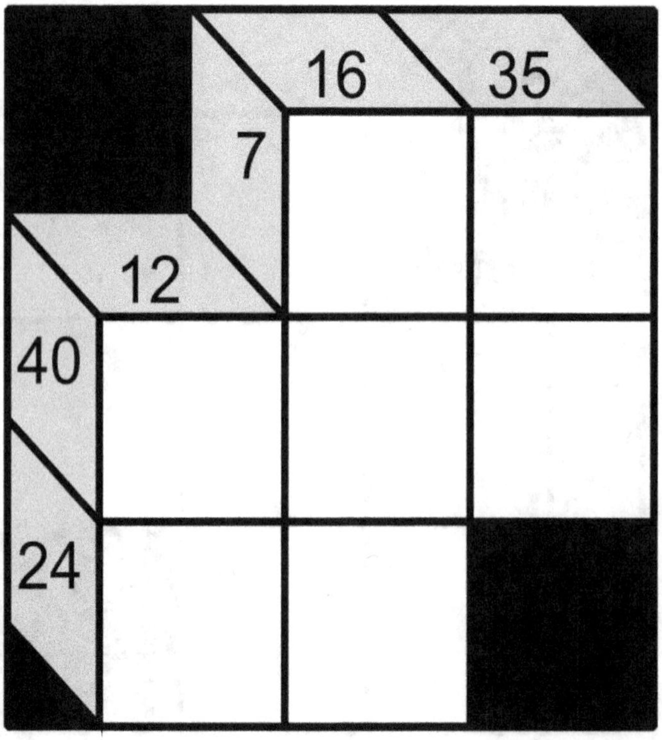

Puzzle 169

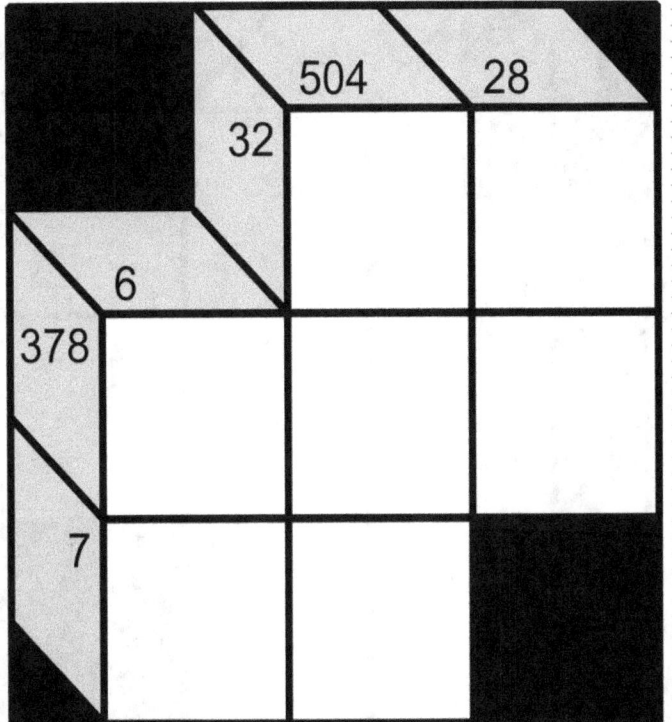

Puzzle 170

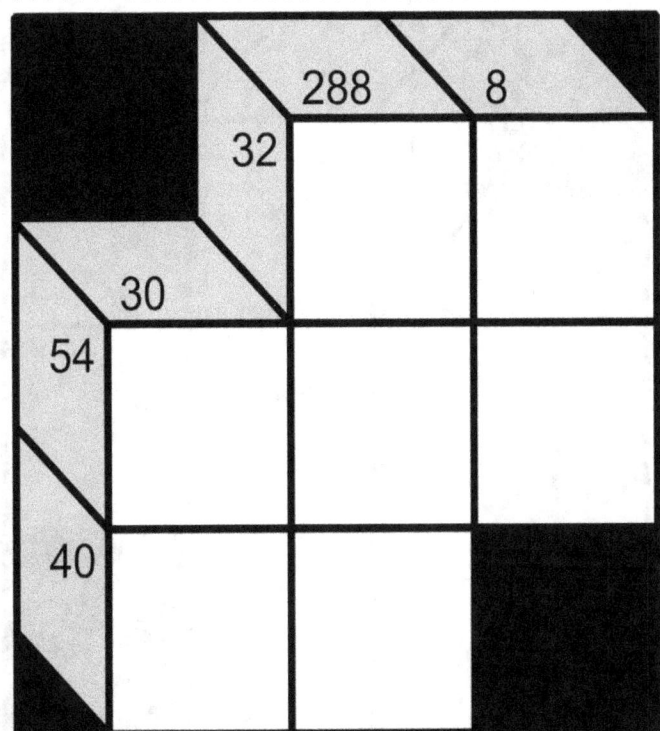

Puzzle 171

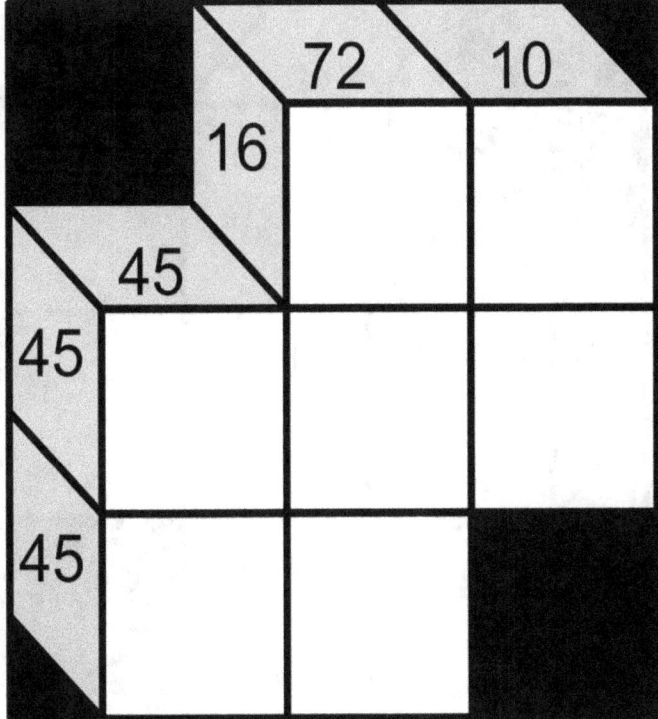

Puzzle 172

Puzzle 173

Puzzle 174

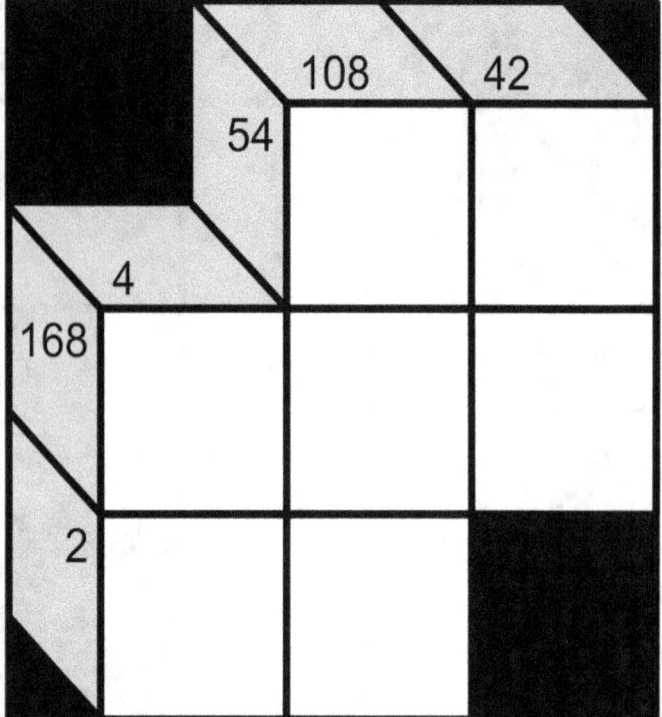

Puzzle 175

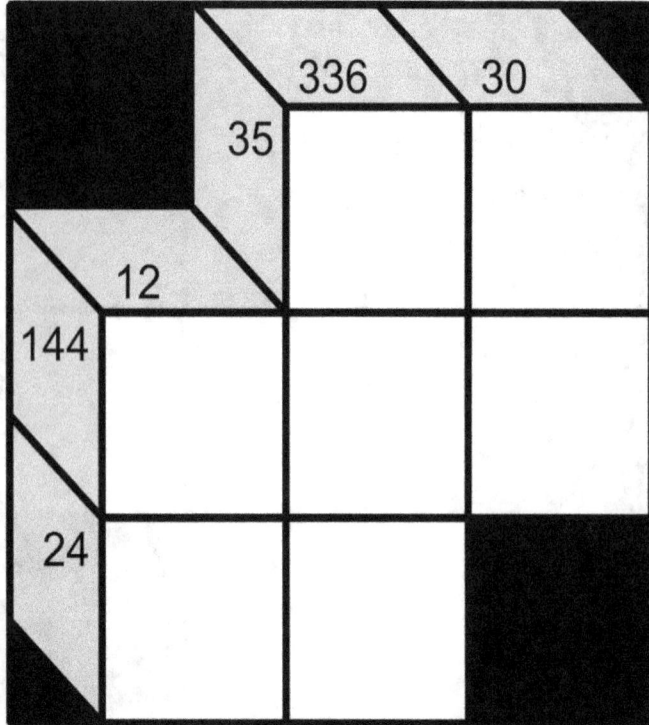

Puzzle 176

Puzzle 177

Puzzle 178

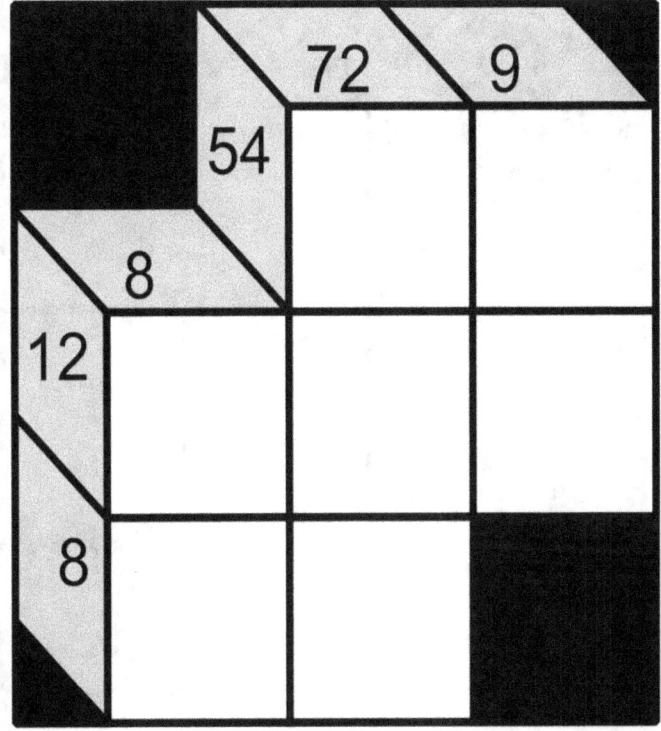

Puzzle 179

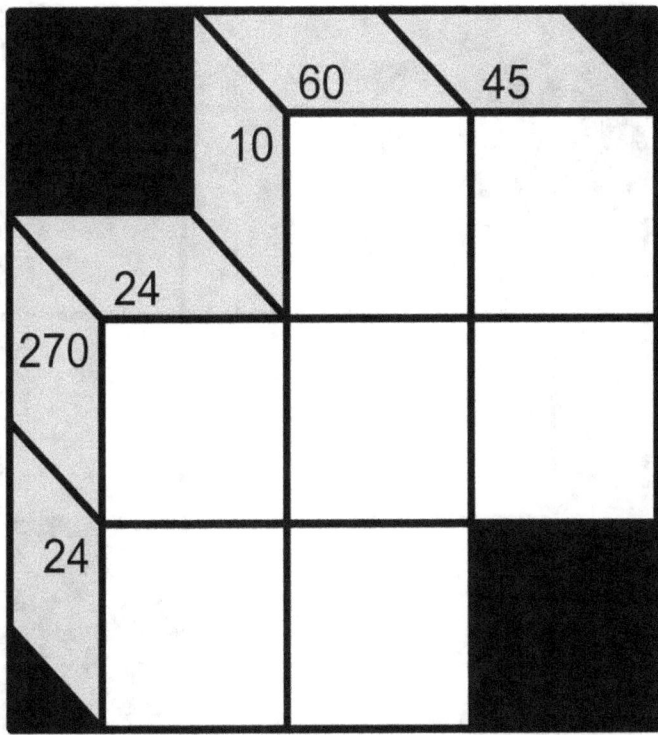

Puzzle 180

Puzzle 181

Puzzle 182

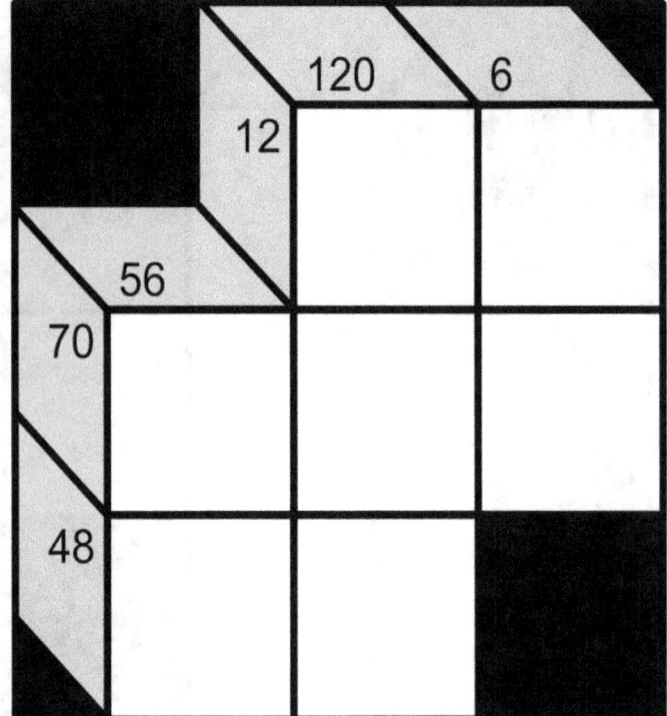

Puzzle 183

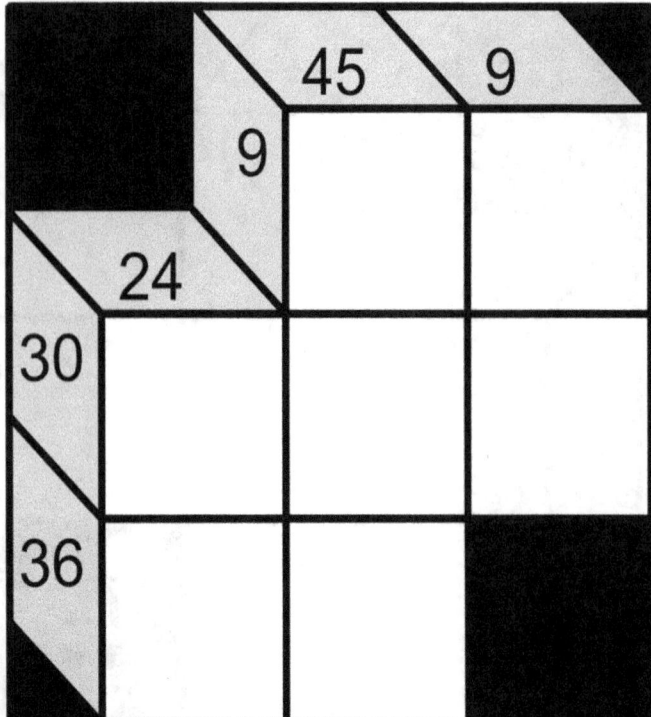

Puzzle 184

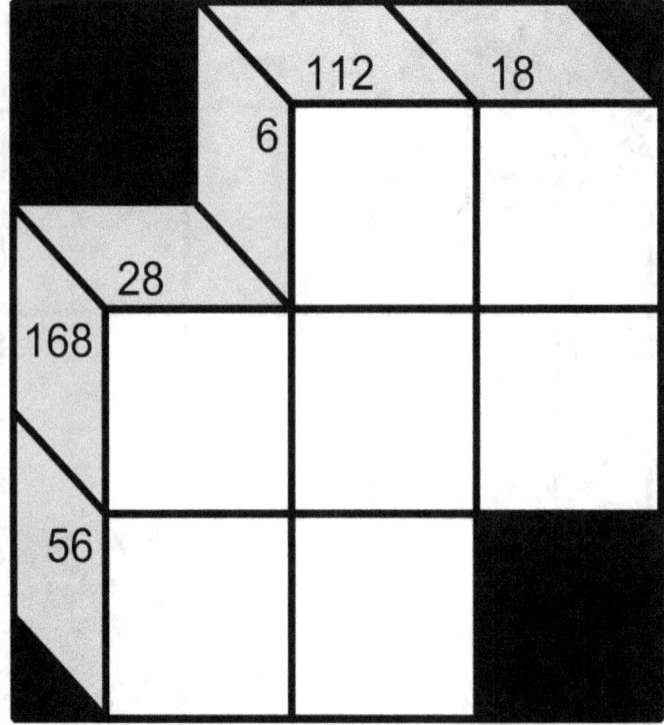

Puzzle 185

Puzzle 186

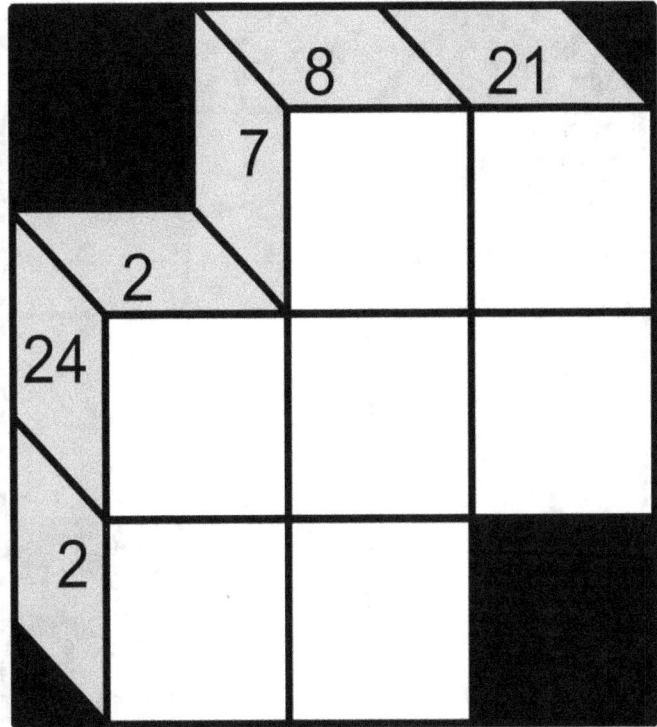

Puzzle 187

Puzzle 188

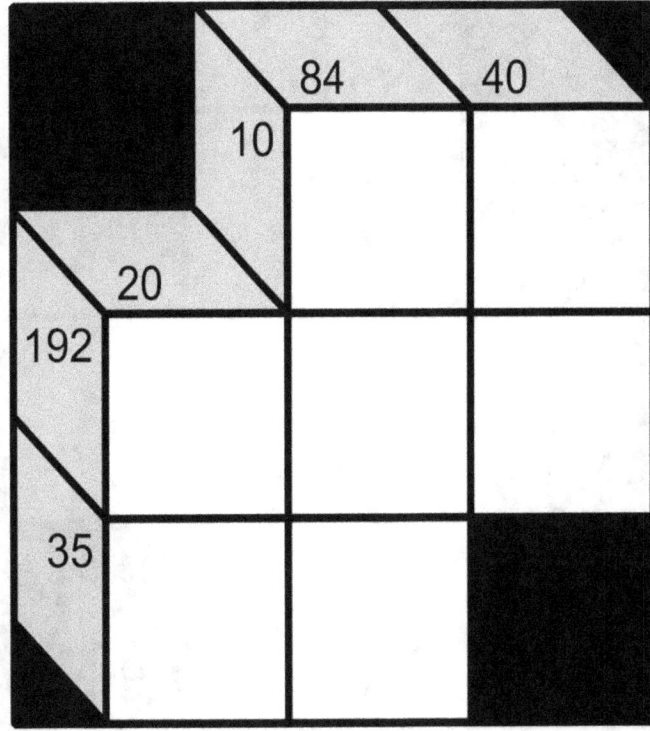

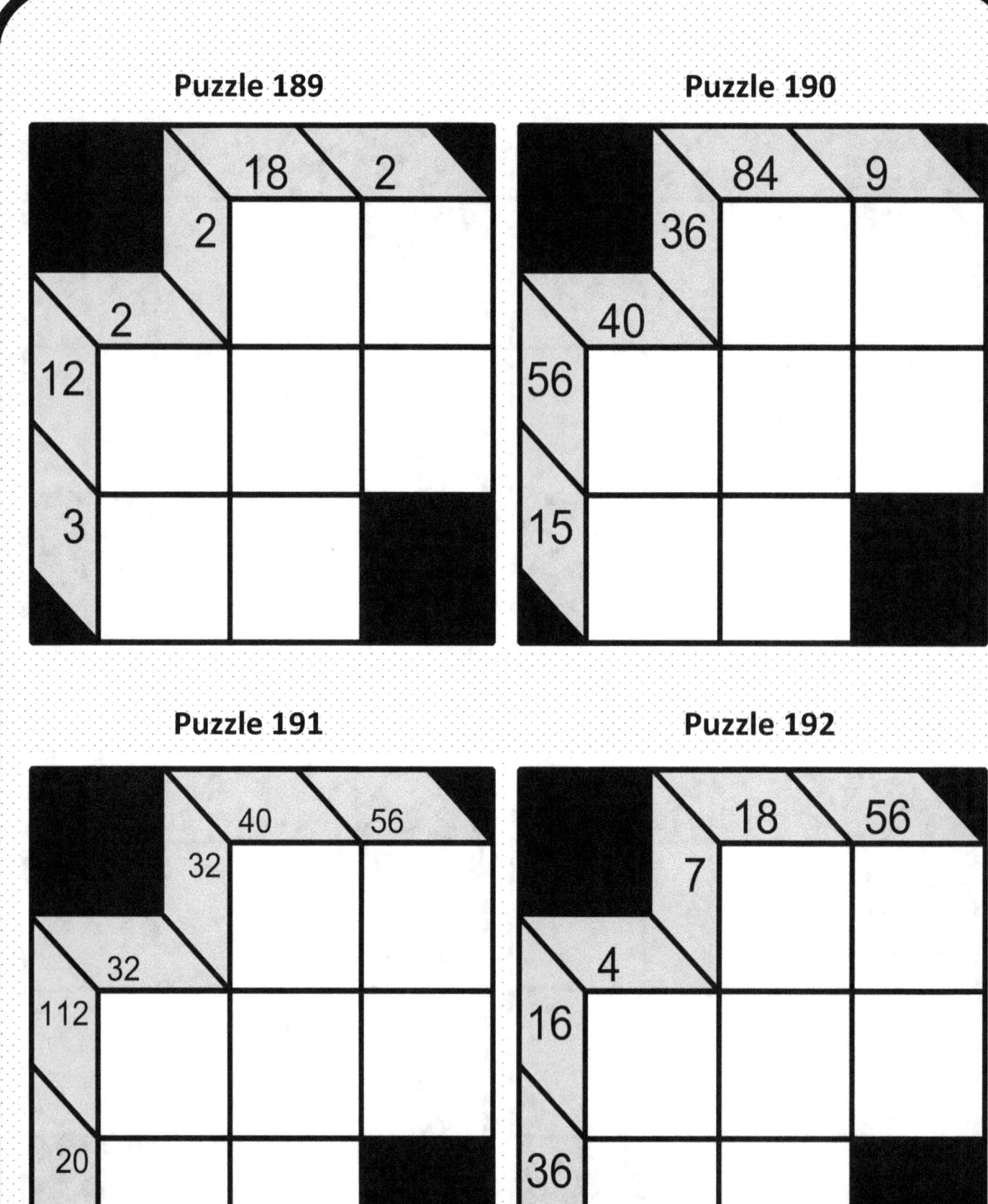

Puzzle 193

Puzzle 194

Puzzle 195

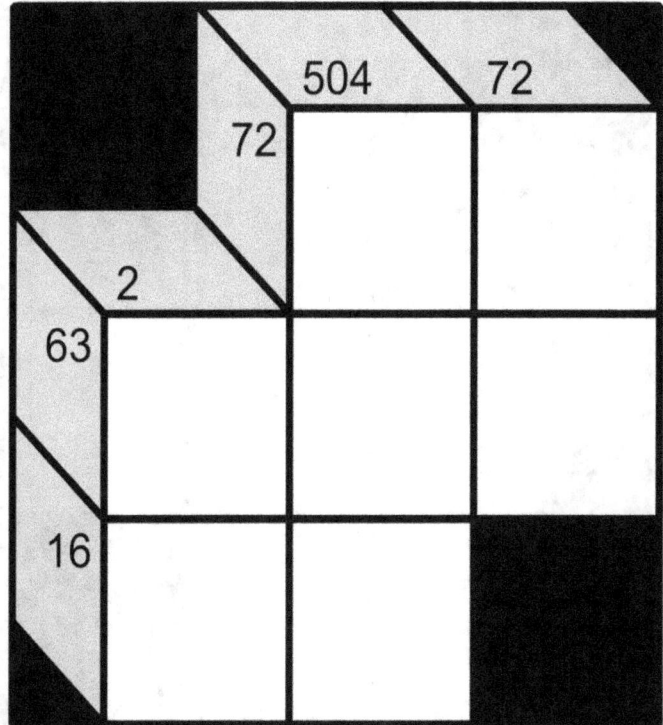

Puzzle 196

Puzzle 197

Puzzle 198

Puzzle 199

Puzzle 200

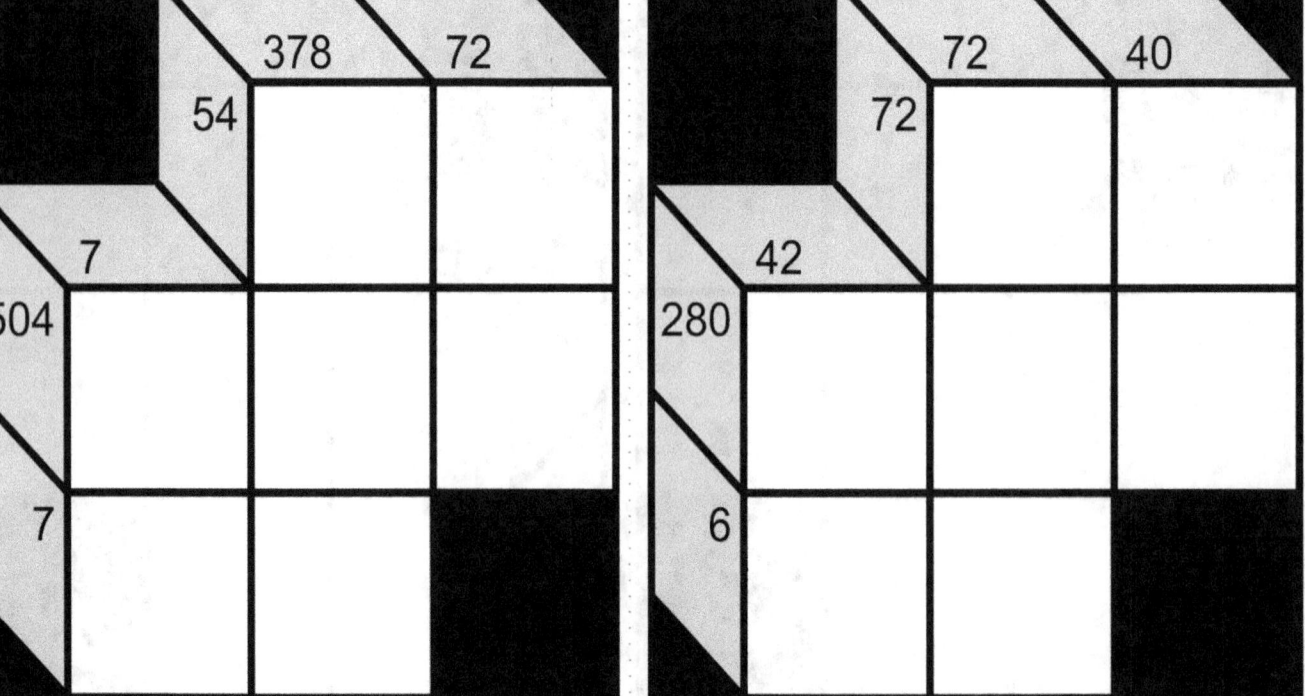

Solutions Start Next Page

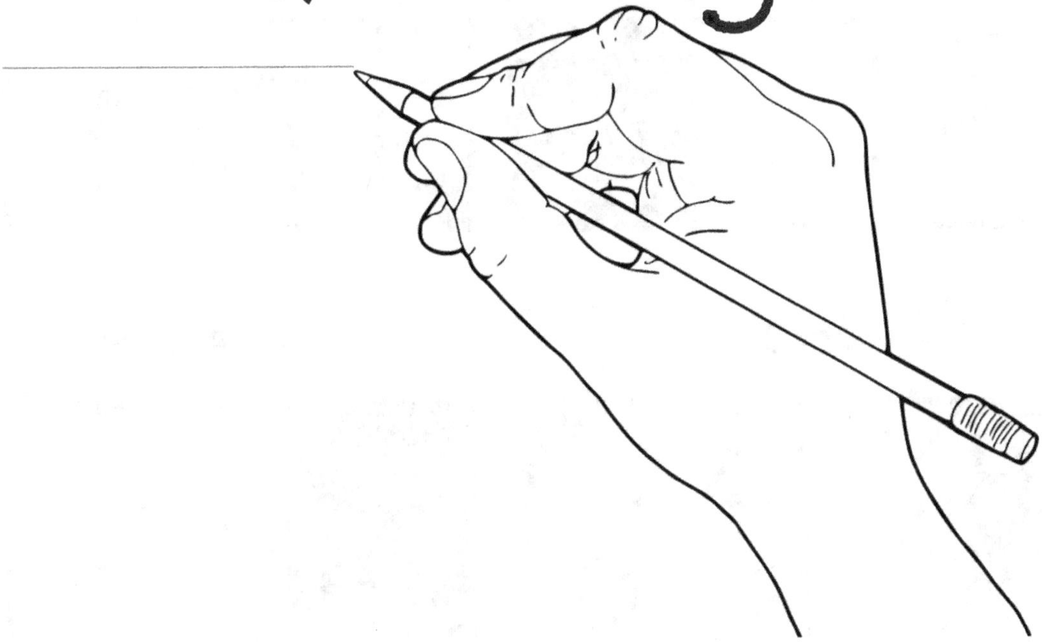

Answer 1

63
7
63
9 7
15
21
3 7 1
5
5 1

Answer 2

12
32
48
6 8
42
48
6 2 4
7
7 1

Answer 3

12
72
54
6 9
4
32
4 1 8
2
1 2

Answer 4

168
8
12
3 4
7
112
7 8 2
7
1 7

Answer 5

	20	18
36	4	9
36 / 18	9	1	2
20	4	5

Answer 6

	96	14
8	4	2
6 / 42	2	3	7
24	3	8

Answer 7

	42	7
21	3	7
54 / 18	9	2	1
42	6	7

Answer 8

	140	15
15	5	3
3 / 60	3	4	5
7	1	7

Answer 9

Answer 10

Answer 11

Answer 12

Answer 13

48
21
56
8
7
20
90
5
6
3
4
4
1

Answer 14

42
27
21
7
3
8
108
2
6
9
4
4
1

Answer 15

15
16
2
1
2
56
168
7
3
8
40
8
5

Answer 16

20
7
7
1
7
54
45
9
5
1
24
6
4

Answer 17

Answer 18

Answer 19
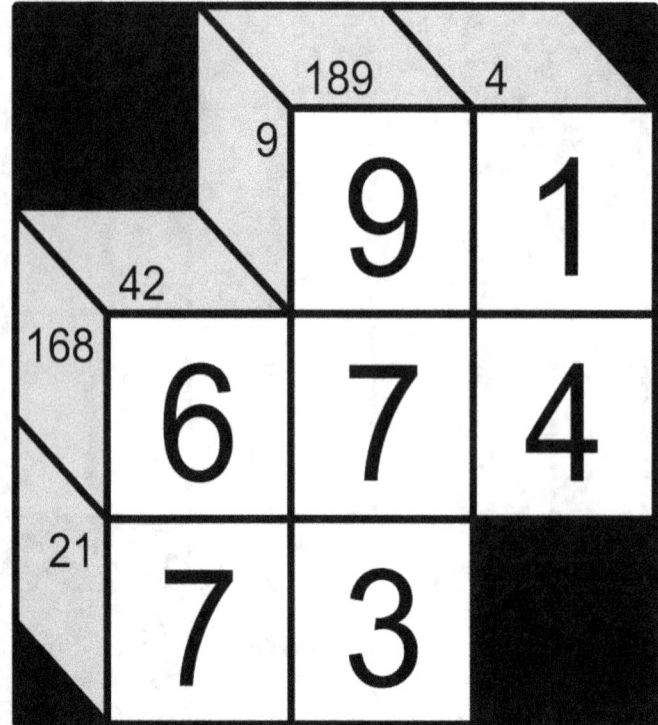

Answer 20
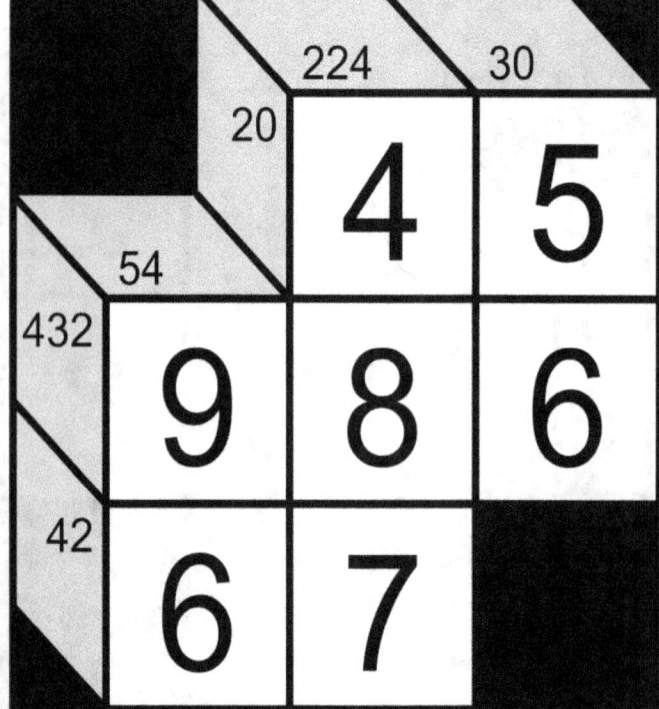

Answer 21

Answer 22

Answer 23

Answer 24

Answer 25

Answer 26

Answer 27

Answer 28

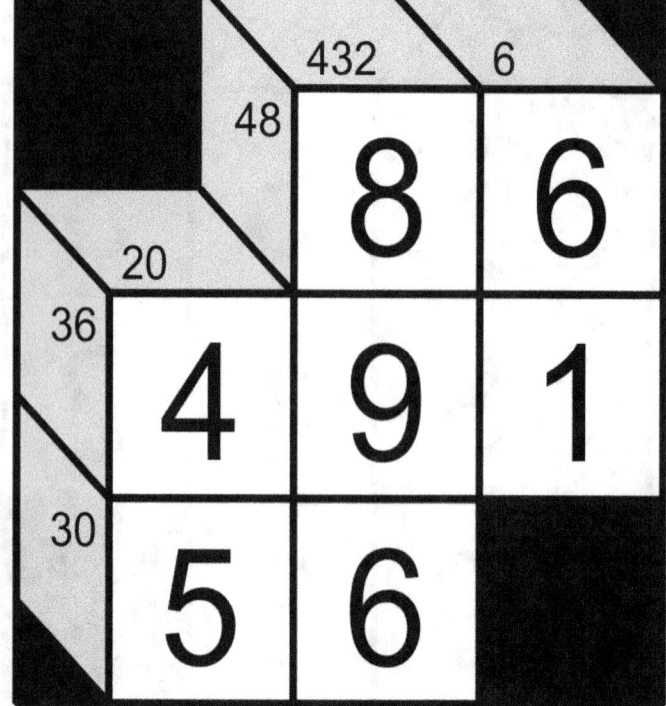

Answer 29

Answer 30

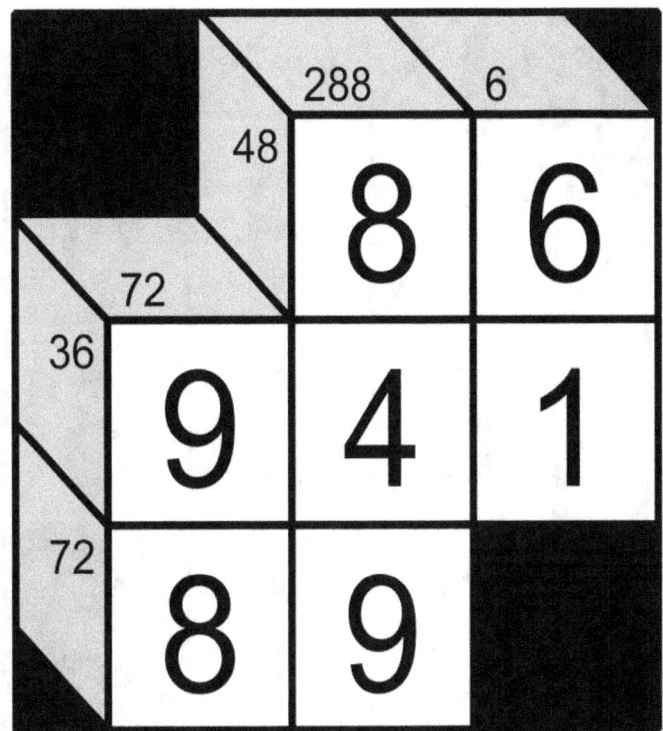

Answer 31

Answer 32

Answer 33

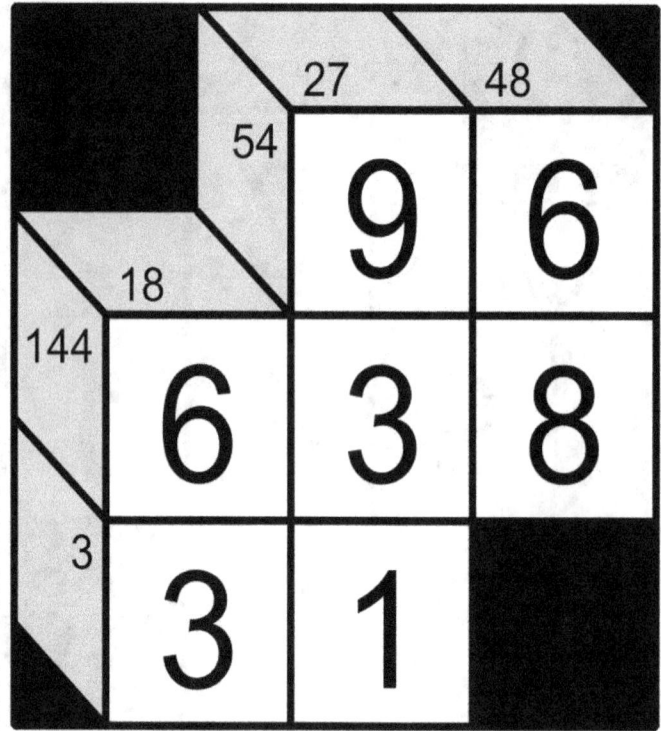

Answer 34

Answer 35

Answer 36

Answer 37

Answer 38

Answer 39

Answer 40

Answer 41

Answer 42

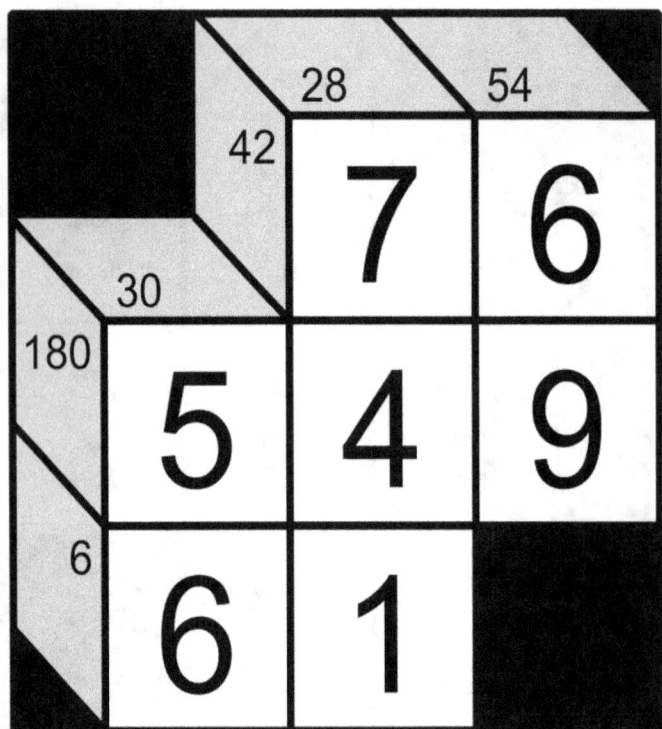

Answer 43

Answer 44

Answer 45

Answer 46

Answer 47

Answer 48

Answer 49

Answer 50

Answer 51

Answer 52

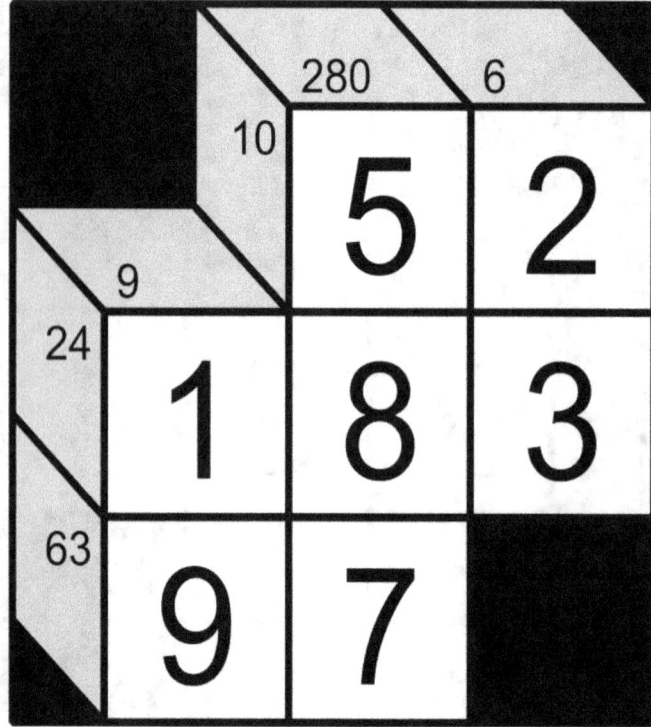

Answer 53

Answer 54

Answer 55

Answer 56

Answer 57

Answer 58

Answer 59

Answer 60

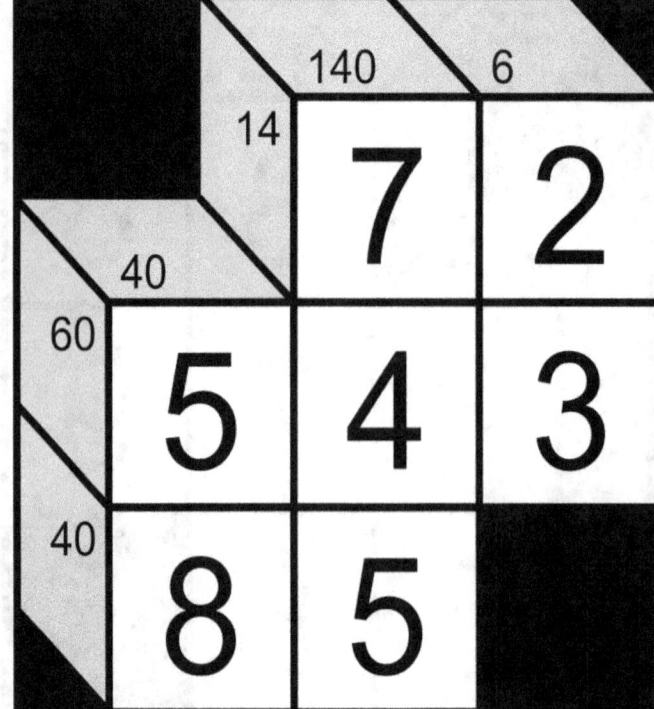

Answer 61

Answer 62

Answer 63

Answer 64

Answer 65

	16	10
10	2	5
45 / 18	9	1	2
40	5	8

Answer 66

	14	24
56	7	8
45 / 15	5	1	3
18	9	2

Answer 67

	20	18
45	5	9
45 / 18	9	1	2
20	5	4

Answer 68

	28	40
20	4	5
15 / 24	3	1	8
35	5	7

Answer 69

Answer 70

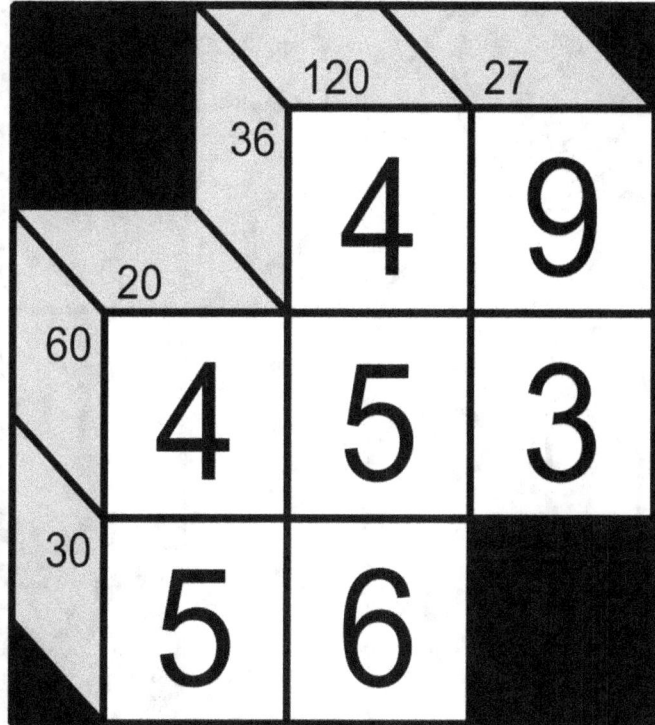

Answer 71

Answer 72

Answer 73

Answer 74

Answer 75

Answer 76

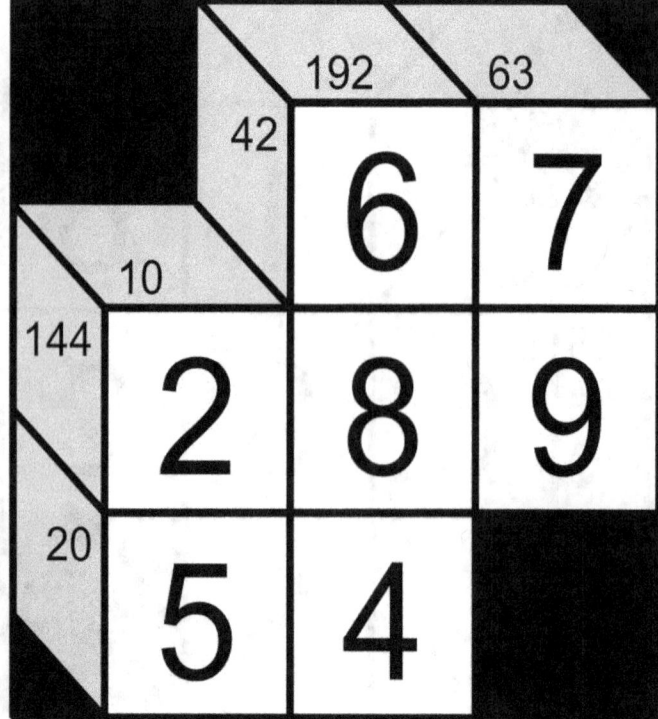

Answer 77

Answer 78

Answer 79

Answer 80

Answer 81

Answer 82

Answer 83

Answer 84

Answer 85

Answer 86

Answer 87

Answer 88

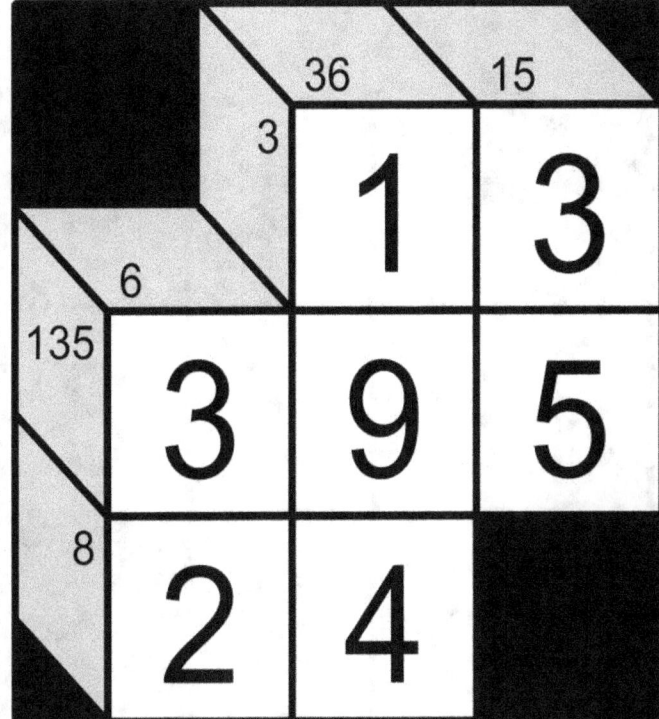

Answer 89

Answer 90

Answer 91

Answer 92

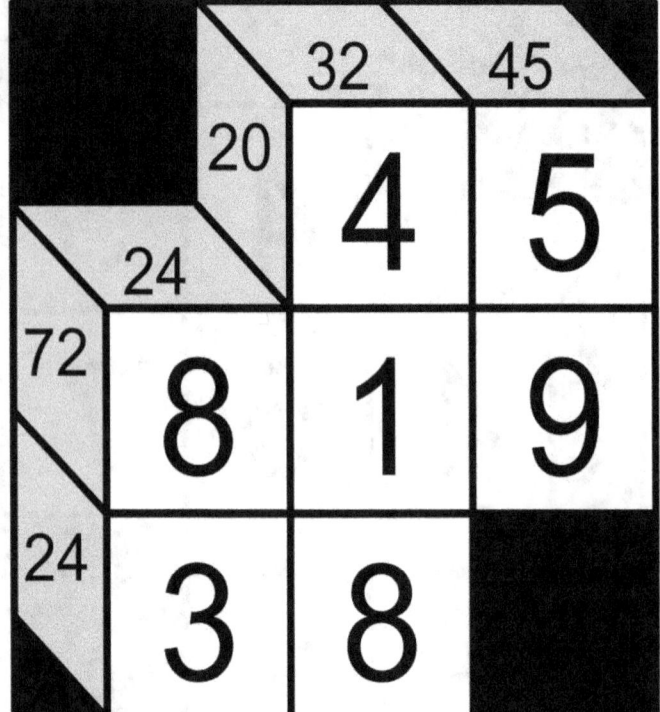

Answer 93

Answer 94

Answer 95

Answer 96

Answer 97

Answer 98

Answer 99

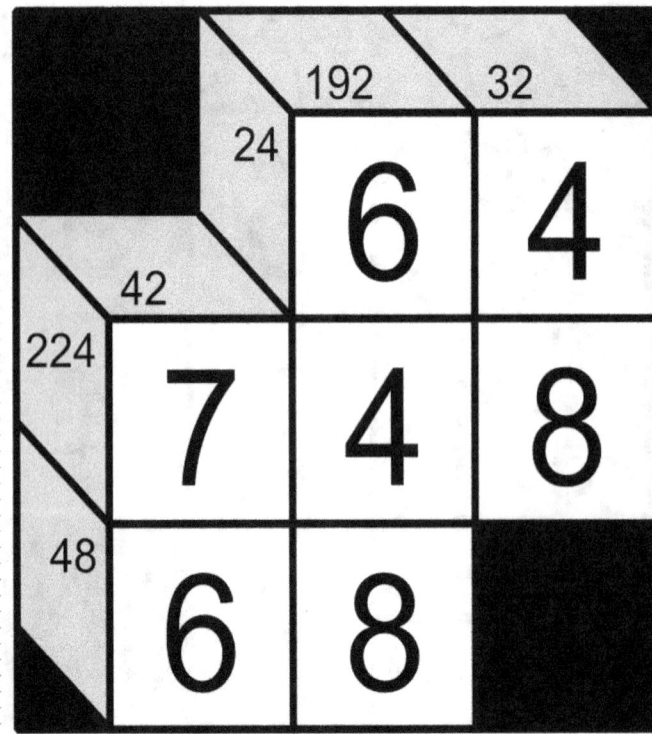

Answer 100

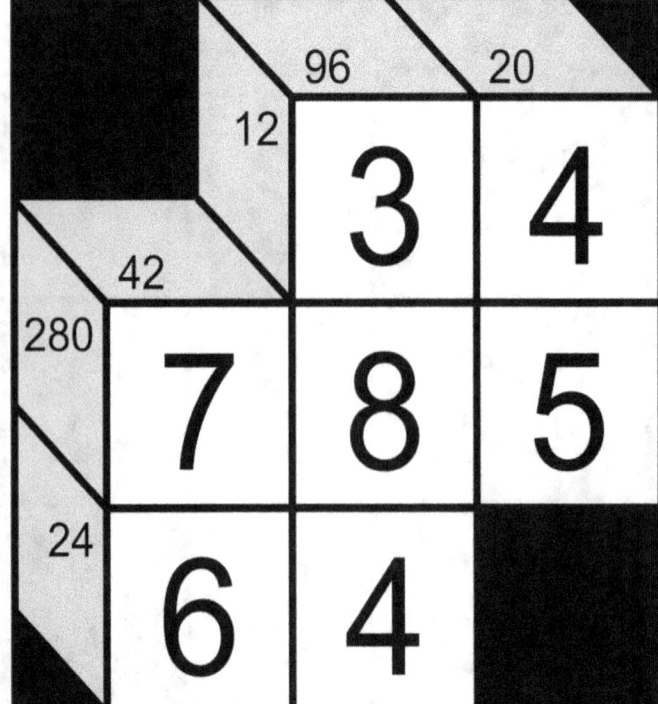

Answer 101

	224	10
20	4	5
2		
14	1	7	2
16	2	8	

Answer 102

	36	4
2	2	1
18		
108	9	3	4
12	2	6	

Answer 103

	108	56
72	9	8
72		
168	8	3	7
36	9	4	

Answer 104

	16	36
18	2	9
56		
224	7	8	4
8	8	1	

Answer 105

```
        54  12
     2
   14    1   2
108
     2   9   6
  42
     7   6
```

Answer 106

```
        56  12
    14
           7   2
    15
240    5   8   6
     3
           3   1
```

Answer 107

```
        12  30
    12
           2   6
    36
120    4   6   5
     9
           9   1
```

Answer 108

```
        72  24
    72
           9   8
    45
216    9   8   3
     5
           5   1
```

Answer 109

Answer 110

Answer 111

Answer 112

Answer 113

Answer 114

Answer 115

Answer 116

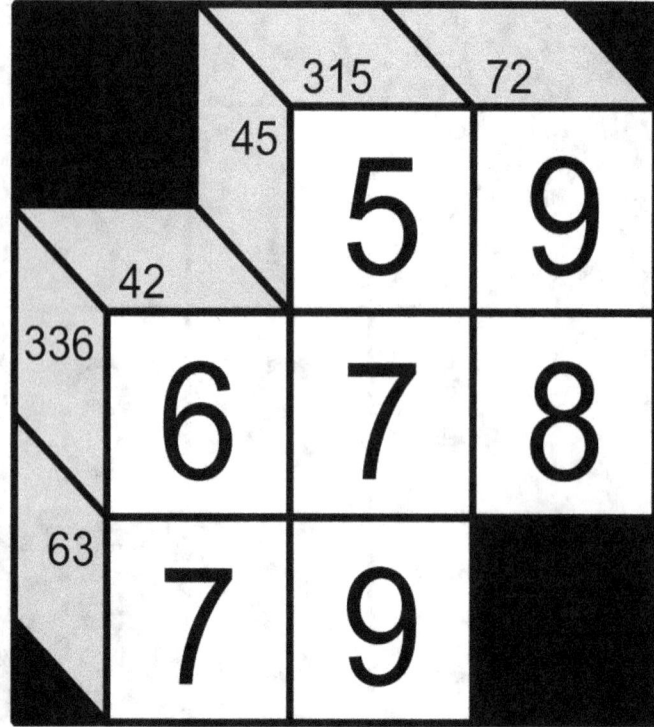

Answer 117

Answer 118

Answer 119

Answer 120

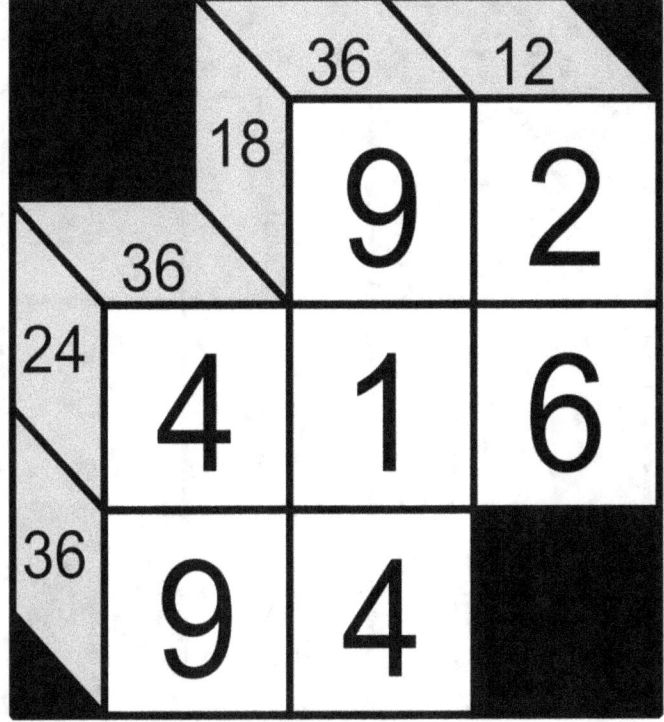

Answer 121

Answer 122

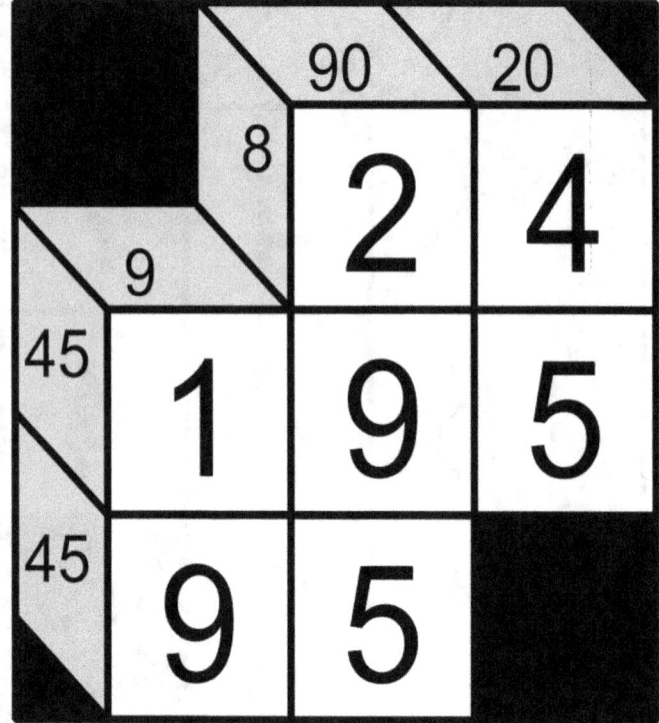

Answer 123

Answer 124

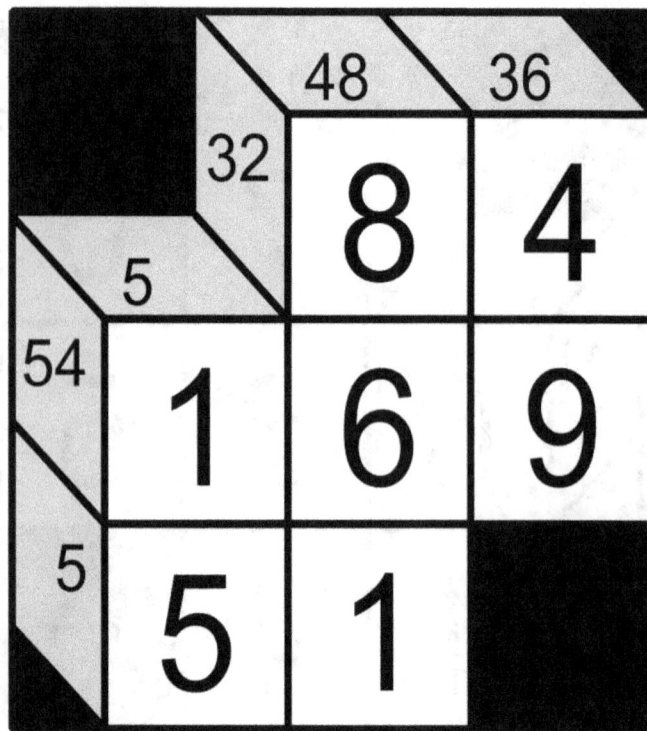

Answer 125

Answer 126

Answer 127

Answer 128

Answer 129

Answer 130

Answer 131

Answer 132

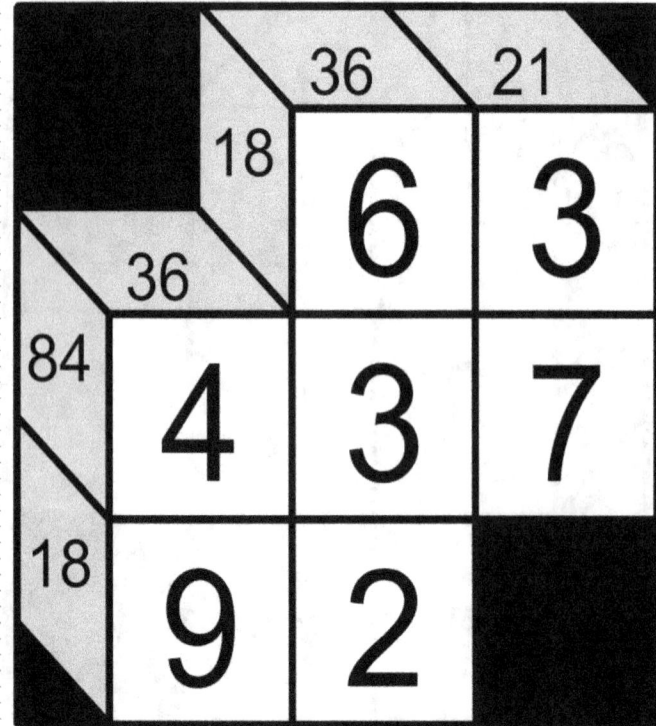

Answer 133

Answer 134

Answer 135

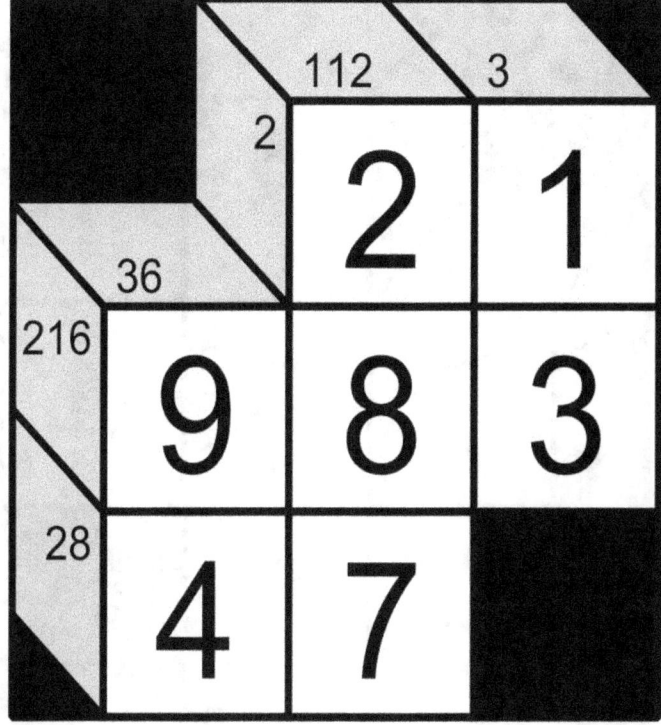

Answer 136

Answer 137

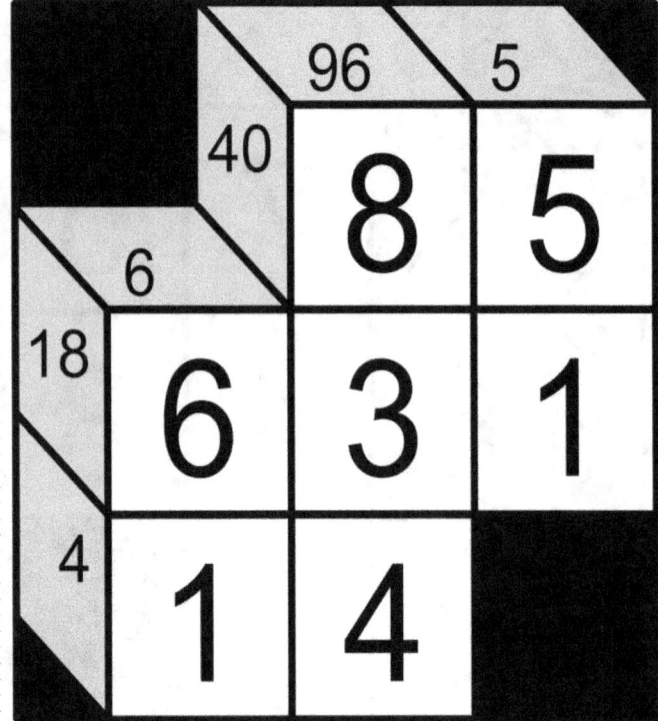

Answer 138

	54	42
7	1	7
28		
216	4	9	6
42	7	6	

Answer 139

Answer 140

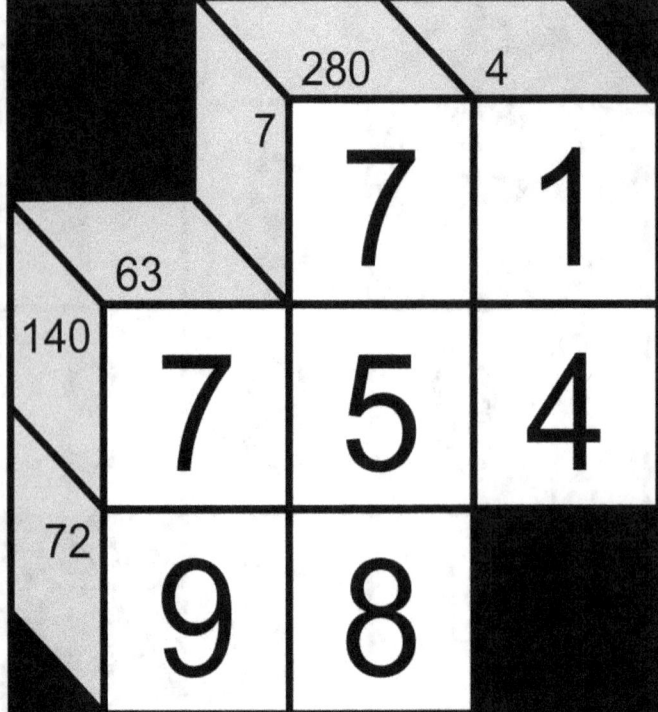

Answer 141

	224	63
63	7	9
15		
140	5	4	7
24	3	8

Answer 142

	18	6
27	9	3
42		
12	6	1	2
14	7	2

Answer 143

	60	18
6	2	3
20		
120	4	5	6
30	5	6

Answer 144

	135	3
27	9	3
36		
45	9	5	1
12	4	3

Answer 145

Answer 146

Answer 147

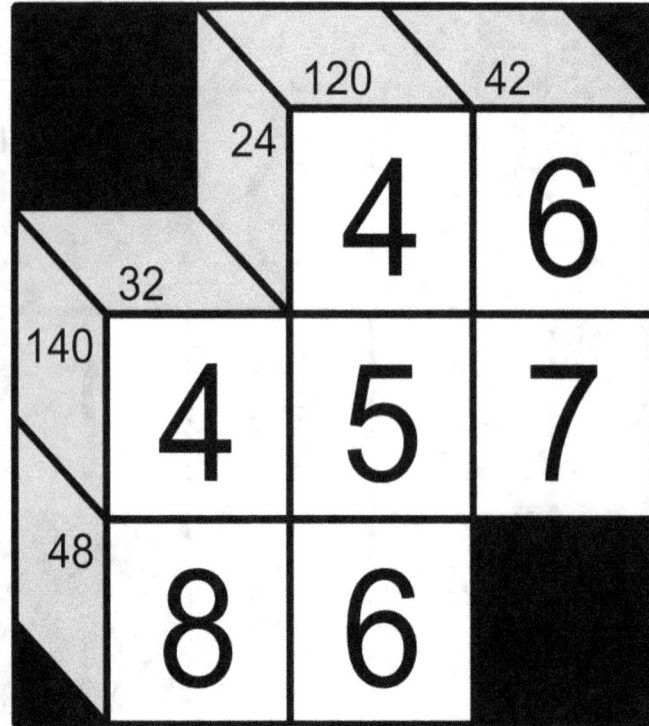

Answer 148

Answer 149

Answer 150

Answer 151

Answer 152

Answer 153

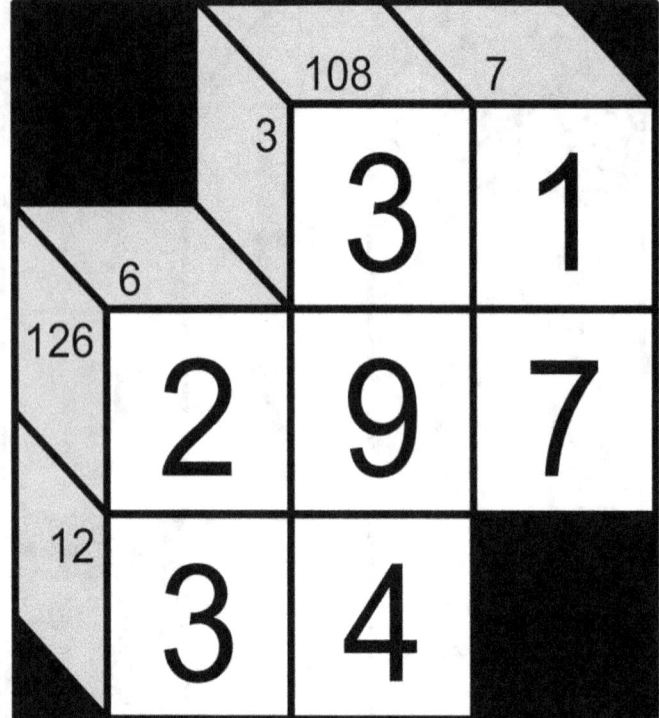

Answer 154

Answer 155

Answer 156

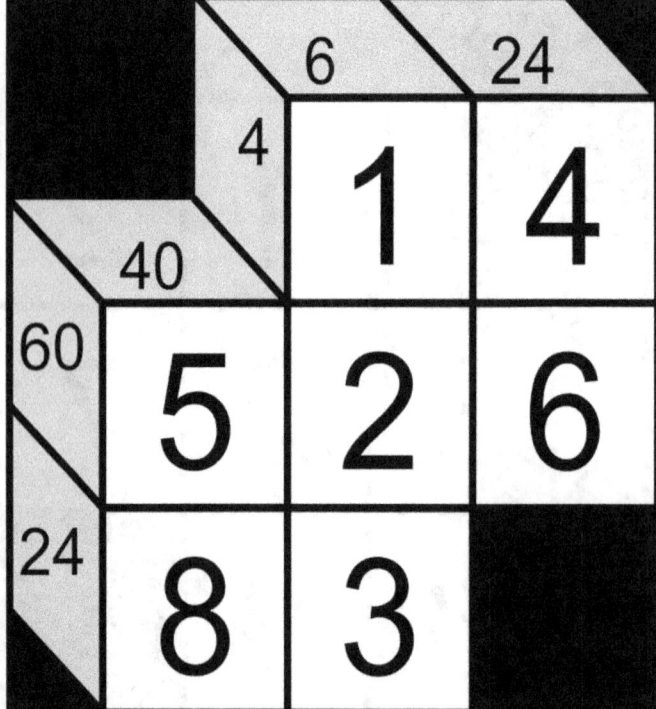

Answer 157

Answer 158

Answer 159

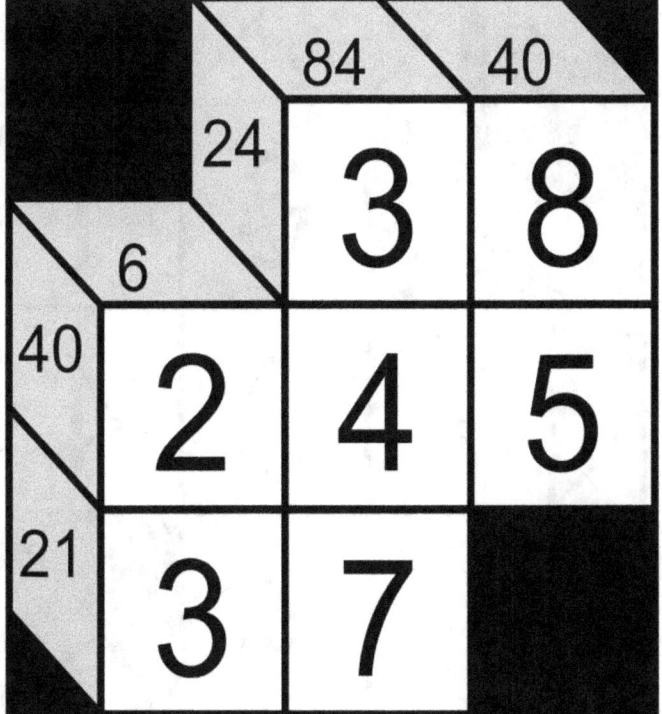

Answer 160

Answer 161

Answer 162

Answer 163

Answer 164

Answer 165

Answer 166

Answer 167

Answer 168

Answer 169

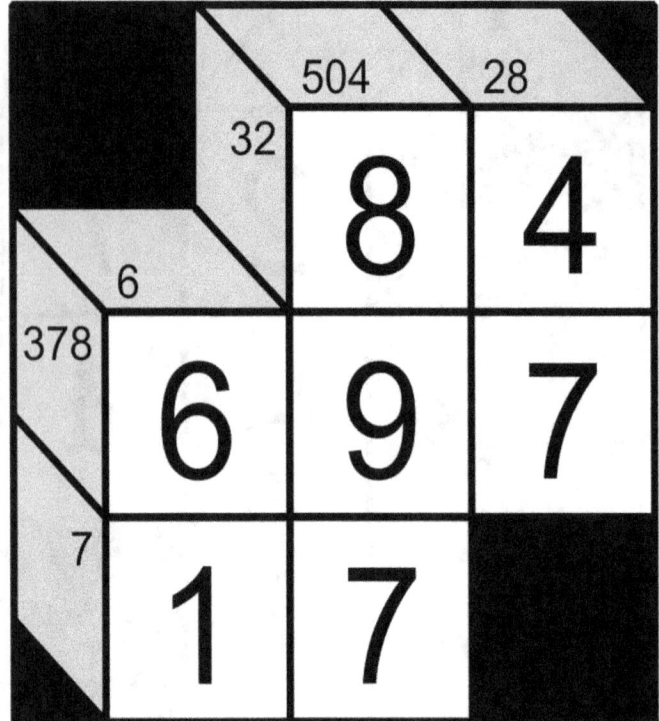

Answer 170

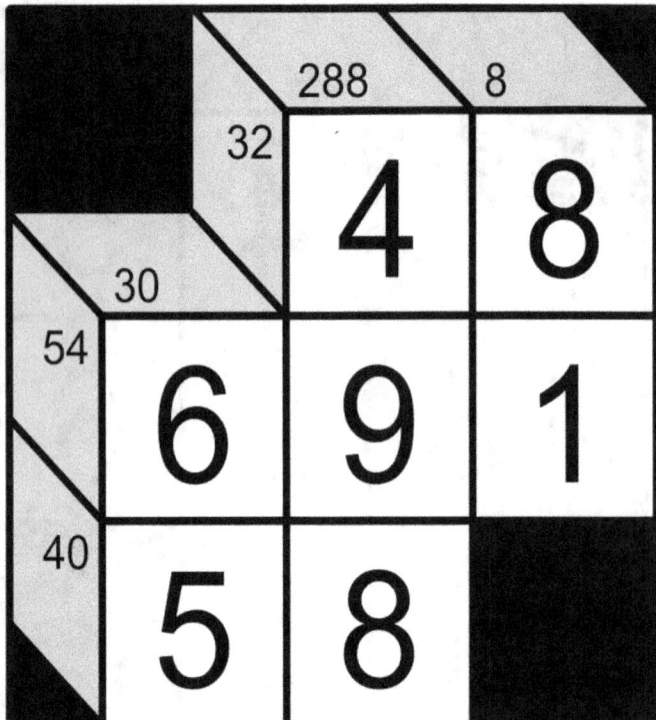

Answer 171

Answer 172

96

Answer 173

Answer 174

Answer 175

Answer 176

Answer 177

Answer 178

Answer 179

Answer 180

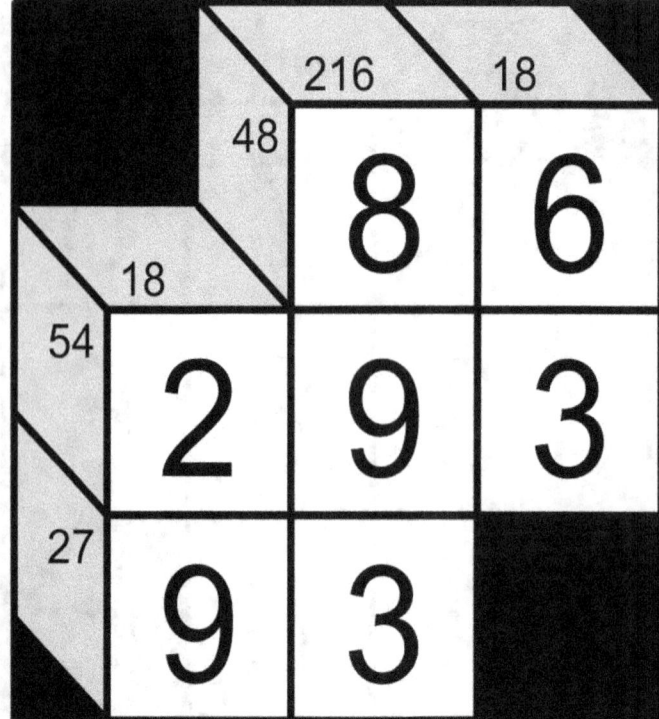

Answer 181

Answer 182

Answer 183

Answer 184

Answer 185

Answer 186

	8	21
7	1	7
2		
24	2	4	3
2	1	2

Answer 187

Answer 188

100

Answer 189

Answer 190

Answer 191

Answer 192

Answer 193

Answer 194

Answer 195

Answer 196

Answer 197

Answer 198

Answer 199

Answer 200

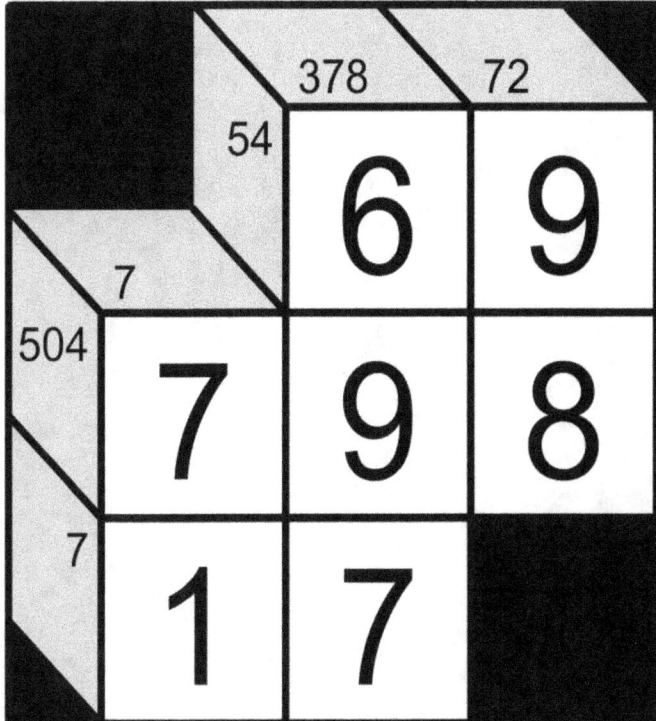

www.ingramcontent.com/pod-product-compliance
Lightning Source LLC
Chambersburg PA
CBHW081443220526
45466CB00008B/2495